Anxiety, Insects, Sugar Cane, and Old Age

Anxiety, Insects, Sugar Cane, and Old Age

Wm. Henry Long, PhD

Copyright © 2006 by Wm. Henry Long, PhD.

Library of Congress Number: 2005908551
ISBN: Hardcover 1-59926-797-7
       Softcover 1-59926-796-9

All rights reserved. No part of this book may be reproduced or transmitted in any form or by any means, electronic or mechanical, including photocopying, recording, or by any information storage and retrieval system, without permission in writing from the copyright owner.

This book was printed in the United States of America.

To order additional copies of this book, contact:
Xlibris Corporation
1-888-795-4274
www.Xlibris.com
Orders@Xlibris.com

Contents

Preface 7
Acknowledgments 9

1. From Whence I Came 11
2. The Early Years (1928-38) 15
3. Farm Life (1938-45) 25
4. Parental and Other Influences 43
5. Wasted Year at College (1945-46) 51
6. Military Life (1946-49) 57
   Basic Training 58
   Overseas Duty 63
   Stateside Duty 78

7. Undergraduate College Studies (1949-52) 84
8. Graduate Studies (1952-57) 96
   North Carolina State University 96
   Love and Marriage 102
   Iowa State University 110
   Birth of a Child 113
   Job Selection 117

9. LSU Faculty Period (1957-65) 120
   The Challenge 122
   Strength from Faith 132
   Contributions and Recognition 135
   Restless for Change 142

10. Teaching and Consulting in Bayou Land (1965-73) 147
    Birth, Death, and Marriage 164

11. Egyptian Mission (1973) 168
12. Midlife Crisis (1974) 197
13. Brazilian Mission (1975-76) 202
    Family Reunited 208
    Mission Completed 235

14. Late Seventies and Early Eighties 243
15. IPM and the Reagan Problem 271
16. The Nineties 287
    University Retirement 293
    Back Surgery 302
    Year of Decision 312
    Final Years of Fieldwork 321

17. Retirement and Retrospection 332
    Better Days 332
    What If 333
    Hall of Fame 334
    Life in Tennessee 335
    Revisiting Nicholls State 342
    Time for Genealogy 342
    Class Reunions 343

18. Eulogy to an Old Friend 345
19. Religious Faith 349
20. Conclusions 367

Summary 377
Author Biography 381

Preface

This story is about early childhood experiences during the Great Depression, rural life on a small Alabama farm, military life at the end of World War II, college years, graduate studies, a troubled marriage, life as a research and teaching professor, the history of integrated pest management (IPM) of sugar cane insects in Louisiana, how a social cripple persevered to help change agricultural pest control practices, the frustrations of working in a small regional university with questionable leadership, the life of an agricultural consultant, missions of an entomology expert to foreign governments of Egypt and Brazil, retirement years, one man's evolution in faith from Protestant to Mormon to nondenominational Christian, and conclusions drawn from a long and confrontational life.

This is the story of my life told as accurately as I can recall and as completely as I dare. I was raised in relative isolation from many of the normal socializing experiences and under a strict puritanical ideology that contributed to an unhappy mixture of anxiety, low self-esteem, frustration, and anger. I take some satisfaction in having contributed to society through science and education while acknowledging serious mistakes and missed opportunities in business, community, and family life. My loving and supportive wife has stood by me now for more than fifty years despite my errors and many shortcomings that have caused serious problems in our home.

Some facts and events have been ignored or deleted from this account to avoid needless embarrassment of others. Some accounts, possibly embarrassing to a few, have been retained at the minimum essential to the accurate depiction of times and situations. May this account of personal history help someone to avoid some consequences of ignorance or poor decisions.

Life has many facets, some very private. I have attempted to fairly summarize my story without dwelling unduly on those less presentable but possibly significant incidents. The only thing we can do about the past is to learn from it. To do so, we must look at it. In this process, we find occasions to both laugh and cry. There were sad times, but there were happy ones too.

Acknowledgments

The inspiration, encouragement, editing assistance, and suggestions on some matters of content provided by my wife's aunt Chloe Hodge were extremely helpful to this effort. I also acknowledge with appreciation the encouragement and suggestions received from other relatives, particularly my brother Needham and my granddaughter Heather, as well as a number of friends who spent time reviewing the manuscript. The editorial staff of the Xlibris Corporation publishers also contributed significantly to this work.

The man on the front cover of this book, pointing to sugarcane borer damage in cane stalks, is Mr. Emile "Max" Concienne, field research assistant for many years to three successive LSU entomologists in charge of research on sugar cane insects. His assistance, experience, and industry enabled me, as a newly employed sugar cane entomologist, who had never before seen sugar cane, to "hit the ground running" in my first week on the job. His contributions to field research in sugar cane entomology have been greater than is generally recognized.

I accept full responsibility for all errors, omissions, or the misinterpretation of facts. If there be errors, they are unintentional. If facts or events have been misinterpreted, they are described as experienced and felt from my own perspective.

These memoirs are dedicated to my wife, Janice, for her many years of love and sacrifice in behalf of me and our children while enduring emotional suffering inflicted by her workaholic husband; my children Janice Faye, Nancy Anne, Daniel Henry, and William Pratt; my grandchildren Holly, Denny, Heather, Heidi, Joshua, Amanda, John, Jake, Dallin, Jillian, and Delaney; my great-grandchildren Gordon, Madeline, and all of those to come; my children-in-law Christina Adams Long, Daniel Stephen, and Robert Williams. May they profit from better understanding one who may have influenced them, little or much, for better or for worse.

This volume is also dedicated to my father and mother whom I love and respect and without whom I would not be here to tell this story, and to my brothers, Robert and Needham, with whom I share a common origin and early environment that have influenced each of us in similar although not identical ways.

1

FROM WHENCE I CAME

My father was born April 15, 1877, on the Long Plantation near Uniontown, Alabama. He studied law independently and was admitted to the Alabama State Bar in 1907. He practiced law for more than fifty years, including four years in the state legislature, a short while as solicitor of the Eighth Judicial District, and thirteen years as judge of the Morgan County Court. As a nearly forty-year-old major of infantry, he served during World War I, volunteering three times for overseas duty but being retained to train troops at Camp McClellan, Alabama. He was an elder, trustee, and scoutmaster, among other things, of the First Presbyterian Church in which he was raised and in which he met his wife. He was a strict disciplinarian with a short temper but a good heart. After his eightieth birthday, he ran for reelection as judge of the county court but was defeated. He was disappointed that the electorate could turn their backs on his years of experience and service in favor of a young "whipper snapper in only his midforties." He then resumed a private practice from which Mother finally prevailed on him to retire at age eighty-five. On the Monday after his death, December 3, 1965, at age eighty-eight, a justice of the Alabama Supreme Court asked at the opening of court that day that the court pause in honor of "a man whose word was his bond," who "was forthright, no one who knew him will deny" and "was a bold, brave sort of man, an all-around good man."

Mother was born October 6, 1900, and died September 22, 1996, two weeks before her ninety-sixth birthday. She was editor of the college newspaper at Converse College in Spartanburg, South Carolina, where she earned the BA degree with high honor in 1922. Years later during summer vacations, she earned the MA degree at Peabody College in Nashville. She taught Latin, history, and art for years in the Decatur Senior High School. She was a very literary person, intimately familiar with classical literature, and able to quote long passages from the Bible, Shakespeare, Tennyson, or you name it. She believed that most worthwhile knowledge could be found in books. This sometimes resulted in humorous behavior, such as parking her first baby for a couple of hours each morning on the porch to be sure that he got plenty of fresh winter air. Perhaps it was the baby's good fortune that the practice was discontinued after a time.

She loved to paint landscapes and flowers and did numerous oils during the last one-third of her life. Some of these were sold; and many hang in the homes of her children, grandchildren, and friends. She loved to play the violin and regretted not having more time to devote to it; she also managed at the piano. She wrote poetry and short stories and sold a few during the Great Depression. To her there was no skill that could not be mastered, no knowledge that could not be assimilated. She believed that determination was all that was required. In her later years, she gave a great deal of her time and substance to a variety of charities. She was involved in prison ministry and was a lifelong member of the First Presbyterian Church, although she preferred to attend Baptist services for a number of years before her death. She loved her children with a passion and was proud of their accomplishments to the point of error. One of my great regrets is that I don't think she experienced sufficient reciprocation of that love from her sons.

My brother Robert observed (perhaps correctly) that, in spite of her keen mind and intellect, she possessed a difficult-to-describe bias against people of wealth and prominence. Although she was honored, loved, and respected by and interacted easily with many such people, she tended to avoid the pursuit of close friendships among them. She had grown up as one among six children in a family with much education and little money. Her mother was a scholar who read Latin, French, and German, was educated by the family governess, loved learning, was fascinated by music and painting, and who, as a minister's wife, taught children in the Sunday school, played the organ for church, and presided over the Ladies' Aid Society while bringing up six children. Mother's father was a Princeton-educated son of a wealthy Indiana lawyer and businessman. He gave away an inherited fortune before marrying her mother and, as a Presbyterian minister, moved his family from town to town and state to state during eight pastorates in four states before retiring. He thoroughly imbued his children with love of the Bible, poetry, and literature and an overwhelming and unnatural fear of even the slightest appearance of "sin" in any of its possible forms. Mother naturally felt that one of her greatest responsibilities was to protect her children from every possible sinful influence, a desirable goal if not overdone.

My brother Needham writes in his book, *Panic: An Odyssey—Anger in Disguise* (Dorrance Publishing Co., Pittsburg, PA, 2005): "My mother's fundamentalism held that if you read the Bible and prayed enough, all of life's needs, developmental included, would be met. The characterization of Jesus and his teachings, which I read in the Bible as a child, were ridiculously incongruous with the distorted, suppressive interpretations that were standards in our home and restricted so many of the simple joys of life. They demanded solemnity, penury, passiveness, self-denial, 'blind' and instantaneous obedience, and strict avoidance of anyone who might be a sinner." While I am somewhat uncomfortable with the harshness of Needham's judgment so freely expressed, there is much truth in it.

For a more detailed analysis and interpretation of the characters and personalities of my parents and how these may have influenced the psychological and sociological development of their children, the reader is referred to my brother's book cited above. My purpose here is more to describe feelings and events of my life, as seen and experienced by me, than to attempt to dissect in scientific detail the reasons for my behavior.

A comprehensive account of my family genealogy was published privately in 1972 by Mother as "The Longs, the Listons, and Some Related Families" by Sara Lapsley Liston Long and dedicated "To Henry, Robert, and Needham." Family members and numerous relatives have copies of this softback volume. In addition to the one hundred seventy-four pages summarizing her extensive research, several Xerox copies of interesting documents are inserted, including an excerpt from the Uniontown, Alabama, *Herald*, December 29, 1887, showing William Henry Long as first corporal in Canebrake Rifles of the Army of the Confederate States, and another from "Records of Land Patents, 1633-1680" in the Hall of Records, Annapolis, Maryland, showing that Morris Liston was a landowner before 1680 in what is now New Castle County, Delaware. The index of Mother's book refers to the following family names: Long, Billingsley, Thompson, Bryan, Bonner, Needham, Liston, Todd, Ashby, Lapsley, Rutherford, Walker, Woods, Armstrong, Poage, Pratt, and Clark.

2

THE EARLY YEARS (1928-38)

At 11:25 PM on September 20, 1928, in the Bank Street Hospital at Decatur, Alabama, William Henry Long III was born to parents, Sara Lapsley Liston and William Henry (Hal) Long Jr. who were married November 17, 1927. Daddy was a vigorous fifty-year-old, twenty-three years older than his young bride. They built a little two-bedroom, single-bath brick house at 226 Fifth Avenue West the first year they were married and moved into it shortly after I was born. Mother described it as "a delightful little house for two grown people and one little boy; but when there were three little boys, it began to be crowded."

I believe that my earliest memories as a child are centered on bedtime and prayers. Mother would sit on the edge of the bed while I recited with her help the children's prayer: "Now I lay me down to sleep. I pray the Lord my soul to keep. If I should die before I wake, I pray the Lord my soul to take. In Jesus name, Amen." Afterward, she would tuck me in and usually move my hands from my groin and fold them on my chest with the admonition that I "shouldn't do that." It worried me to hear that I shouldn't do that and my usual response may have been among my earliest acts of rebellion.

After the year of my birth began the Great Depression that was the worst and longest economic collapse in the modern industrial world, lasting from the end of 1929 until the early 1940s. I can still remember having cabbage and grits, it seemed, at every meal during the 1930s. Daddy had a private law practice at that time, but few people had enough money to pay for the bare necessities, much less legal fees. Farmers peddling their products door to door would occasionally leave a bag of produce, a chicken, or part of a butchered animal in partial payment of a legal fee. I clearly remember seeing fresh hog brains for the first time in a bucket and hearing Mother cheerfully proclaiming how lucky we were to feast on brains and eggs for breakfast that day. Of course, we were fortunate then to have so much protein for breakfast. However, I've never since eaten brains and eggs. Mother claimed there was one whole year when less than a dollar passed through her hands.

Their second son, Robert, was born in 1930 and their third, Needham, in 1932. Mother saw to it that all three of her boys were reading and writing before they ever saw the inside of a school building. Those skills were learned at an early age as was the art of tying shoes. I remember the three of us boys sitting on the front doorstep of the house on Fifth Avenue West practicing tying pairs of outgrown or worn-out shoes that had not yet been disposed of. This was a routine performed daily until we were proficient at tying shoes. After a short break, if you could hold a pencil, you would work at printing and writing letters of the alphabet. These activities would later evolve to conjugating verbs and writing declensions and inflections of nouns, pronouns, and adjectives. In these respects, we were well ahead of the average children our ages.

Christmas was a truly exciting event for us children, though much less extravagant by today's standards. A day or two before Christmas Day, the family would participate in finding and cutting a cedar tree to be placed in a bucket of stones to hold it upright. Old-fashioned tinsel icicles and a few colored glass balls in addition to homemade decorations of crayon-colored paper chains and popcorn strings commonly were placed by little hands before adjustment by more mature artistic ones. Stockings were hung to be filled with apples, oranges, and peppermint sticks. Daddy usually made a quick trip to Kuhn's five- and ten-cent store after supper on Christmas Eve to shop for Santa. However, the real Santa was Aunt Mae, Daddy's spinster sister, who always arrived by train from Birmingham, where she worked for years as senior legal secretary for the law firm of Bradley, Baldwin, All, & White. She and Daddy were the only surviving members of their family, so we were the only family that Aunt Mae had. We always looked forward to Aunt Mae's gifts, which sometimes included such extravagant items as toy trucks, motorcycles, and lead soldiers. And one Christmas, there was even an electric train with passenger cars, freight cars, and caboose.

Though always cordial to each other, as far as I could tell, Mother and Aunt Mae were not the best of friends. I think that Mother resented Aunt Mae's intrusion into our family on these rare and special occasions like Christmas and Thanksgiving. Aunt Mae always seemed to have the means to bring special gifts and remembrances for special occasions. Such was not commonly possible in our home. Aunt Mae always dressed fashionably and wore jewelry and lipstick. Mother was always attractive, clean, and neat but never had a large wardrobe or even wanted one apparently; she did not care for jewelry, wore little if any makeup and only clear or very pale lipstick. Aunt Mae had difficulty concealing her prejudice for her favorite nephew, Henry, whose gifts seemed sometimes more expensive than the others. The reason for this may only have been that I was her brother's namesake. Nevertheless, Henry was the only one to ever ride the train to Birmingham for a weekend with Aunt Mae in her two-room apartment, with Sunday dinner at a fancy restaurant plus a tour of Birmingham and the Vulcan statue, all hosted by the senior member of her law firm.

There was no time or money for family vacations, and our entire family never traveled out of town for a vacation together. However, there was a summer, in about 1934, when a friend of Mother's, who owned rental property on the beach at Mobile Bay in Fairhope, Alabama, invited us to come and stay as her guest for a week in one of her cabins. Daddy was too busy to go, but Mother took her three boys, aged six, four, and two, to the beach. We traveled there and back by train on the L&N Railroad. The cabin at the beach was comfortable by 1930s' standards. There were fans to move the air and speed the evaporation of body sweat that eventually permitted blessed sleep in spite of the sounds of mosquito wings. None of her boys could swim yet, and I, being the oldest, needed to learn. Therefore, our scheduled daily activities included an hour by the clock during which Mother would stand in water to her waist with a box of cookies while urging little Henry to jump from a pier to swim at least one stroke for a cookie. I ate a few cookies that week but did not yet learn to swim. A stray puppy, left on our porch, rode the train home with us in a box under the seat. The conductor thought he had heard an illegal bark but decided he must have been mistaken. After a week back home, we boys all had ringworm, as well as the pup which gave it to us.

Because Daddy was fifty years old when he married for the first and only time, he and Mother were eager to get the boys through school as early as possible to reduce the chances of her being left a widow with three children to educate. Therefore, all of us were more than a year younger than our classmates. At that age, even one year was a significant difference. It meant that you generally could not compete on an equal playing field with your associates who were older, more experienced, and stronger than you. While success breeds success, defeat breeds defeat in a continuing cycle that undermines self-esteem.

Mother believed that movies generally were an evil influence with a few exceptions such as "Snow White and the Seven

Dwarfs" and "Sound of Music." Of course, there was no money available for such activities at that time anyway. However, when we occasionally were invited by a neighbor, Mrs. Burleson, to attend an afternoon movie with her son D. D., who was a year older than I, we usually declined such invitations. It was due not only to lack of money; with Mother, it was also a matter of principle.

We might have benefited by having a sister that we never had. As we became more aware of girls, we also became uneasy around them. We seldom played with them, and those at school were noticeably older. Girls were different, and it wasn't nice to be obviously curious or much concerned about human anatomical details. However, on one occasion, a little six-year-old female friend of my brother Needham from down the street was visiting while I was in bed with some forgotten malady. They were crawling about the room and eventually crawled under the bed. Mother came in to find where they were. When they were found comparing each other's anatomy, there was punishment and reprimand for both. Thank goodness, I had not looked! In describing this event, my brother says that it was dark, and he never saw what he was presumed to have seen anyway. However, we both agree that our mother's reaction to this and other things was inclined to make us feel that sex was bad, wrong, and sinful. May God rest her soul; she was only doing her best to imbue us with the same values that had been thoroughly drilled into her by her saintly father.

In social situations, we were taught as children to "be seen and not heard." This was particularly true at the dinner table. A vocal outburst of any sort or crying could send you to a dark closet until you got control of yourself. Also, all food put on our plates was to be eaten and not wasted. Failure to eat everything on your plate commonly required that you be sent to bed to recover from whatever illness was ailing you. To this day, I feel obliged to eat every particle of food on my plate whether I am at home or in a restaurant. Corporal punishment was frequent enough to maintain continuous fear of Daddy with his razor strap or, more rarely, Mother with a switch from the yard and commonly severe enough to raise whelps that could be seen for a day or so.

At age nine, my feet were almost as large as Daddy's, and at age twelve, I would wear an adult size 12 shoe. Whenever we shopped for shoes, which was usually at Wohl's Department Store, Mr. Wohl himself would wait on Daddy and always exclaim: "My god, Hal, where did that boy get such big feet? Nobody will be able to fit him in another year or two." I always knew what was coming before Mr. Wohl appeared and wished there was some way to get it over with as quickly as possible. I believe that in this matter, Daddy's feelings were similar to my own.

An exciting incident from the early years, which I remember rather clearly, was a mud ball battle in which mud balls were hurled back and forth between two armies. My brothers and I were the good guys who were pitted against the enemy that included the rest of the neighborhood. The enemies were our next-door neighbor, D. D. Burleson; the dentist's son from across the street, Ned Anderson; another older boy and instigator from across the street, Morris Jones; and the son of a Nehi bottled drinks distributor from down the hill, Thomas Craig. There may have been two or three more enemies; I'm not certain. However, the battle line was drawn up in our front yard with the Long brothers' backs immediately before the front brick wall of their home while the two younger brothers made mud balls as fast as possible for the older one to fire at the enemy as rapidly as possible. At the time, Mother had been helping Daddy at the office with some typing. The battle ended as their car appeared coming up the hill. To my surprise and satisfaction, we were only scolded and required to hose the mud off the house. I think Daddy was pleased to see his sons standing together so bravely in battle.

Mr. Burleson and D. D. often went fishing in their boat and brought back good catches that we could see and hear them cleaning and talking about over the back fence. Dry Creek canal, not a popular fishing spot, was about three blocks from our house. It is still there and has had water in it all the three or four times that I have seen it during the last sixty years. I desperately wanted to fish in Dry Creek but had no fishhooks, and there was no money for fishhooks. Imbued with little more than the philosophy that determination overcomes all and with a suggestion from Mother that I might fashion a fishhook from one of her straight pins, I proceeded to do so. Needless to say, I couldn't keep a worm on the hook or the hook on the line long enough for a fish to strike.

When I was eight or nine years old, I determined to run away. Life at home seemed unbearable, and I just wanted to go away to anywhere. I believe that I contemplated it for several days and nights until one day about midmorning, after Daddy had gone to work and Mother was busy in the house, I put a few things in a paper bag and eased out the back door unnoticed. As I passed old Snow, Daddy's setter bird dog, I reached through the fence to pet him goodbye. Then I was out the back gate and walking slowly down the alley toward Dry Creek. The farther I went from home, the more slowly I walked. I turned and walked up Fourth Avenue and back down the other side of that street to get one last look at the park, the swimming pool, and the neighborhood. Then I ambled on toward the Dry Creek Bridge. I was standing on the bridge looking at the water when a car pulled up beside me and Daddy told me to get in, which I did immediately. I do not remember being physically punished for this. However, I was thoroughly scolded by both parents as most children might have been under similar circumstances.

In fourth grade at Lafayette Street School, Duby was a tough large teenager from a tough part of town. His desk was in the first row of Ms. Spear's class. All the desks in that row were noticeably larger than those in the other rows. They were occupied by boys who were noticeably larger than the other students. There was no automatic social promotion in those days. Two or three of the boys in that row boxed at the Armory on Friday nights where people paid to watch. Duby was not a boxer, just an overweight kid with a low IQ who loved to take out his frustrations on average fourth graders playing marbles during recess and lunch periods. "Sweepers keepers, losers weepers" was one of his favorite games. In those days, I would never have dared to stand up against Duby or to tattle on him either. Watching out for Duby was like being a chicken watching for the hawk. No fourth-grade boy of normal age could stand up successfully to Duby.

Joe was no Duby, but he seemed to be a little older and tougher than the average fourth-grade boy. He enjoyed intimidating kids whom he selected for that purpose. I was one of the kids on his list in fourth grade. I was age eight going on nine, and he was probably ten or more. One of his favorite pastimes was to follow a few steps behind me from the school yard to Daddy's office one block away while spouting threats and verbal abuse. When I finally reached the stairs to Daddy's office beside the Roxy Theatre, he ceased following, and I found blessed refuge. I was ashamed for my father to know that I was afraid of this boy. Without telling him anything, I would pretend to do homework until time to go home while fantasizing about whipping Joe.

Jean Link was a beautiful blonde blue-eyed girl, a member of our church, and in my sixth-grade class at school. She was at Malone's swimming pool one spring day with her mother who watched from the bleachers. My mother had permitted me to take a break from painting in the little house on Fifth Avenue West that we were preparing to sell before moving to the country. The swimming pool was in the park three blocks from our house, and I had permission to go there for an hour. Encountering Jean Link at the pool was both exciting and frightening. So few people were there that it was difficult to pretend I didn't see her, and she was so beautiful that any positive experience with her I thought could be only fantasy. She smiled at me and swam to my ladder to talk for a few moments. I was flattered but had no idea what to say to such a gorgeous creature. In less than a minute, she was gone, and her space was occupied by Maurice, a neighborhood bully about my size but perhaps a year older. He immediately spewed a mouthful of water in my face and in a daring tone said: "What you gonna do about that?" I bravely replied, for the benefit of my audience, that he had better not do it again. Of course, his immediate response was to do it again. Wishing to impress my audience without making a scene, I told him to meet me behind my house and left the pool to resume painting walls. Shortly thereafter from my stepladder, I heard the sounds of someone calling and whistling from the alley. With shaky hands and enough adrenalin flowing to pull a truck, I went to the alley to bob, weave, and roll on the ground with Maurice. I got in a few lucky licks before we parted, both too exhausted to continue. Maurice came to school the next day with a black eye. I was publicly humble but privately ecstatic over my first significant victory in physical combat.

Before we moved to a farm in 1939, our parents owned another small farm west of Decatur on Trinity Mountain. There were only a few cleared acres of highly eroded red clay there with many gullies surrounded by woods. A small house and barn were on the front of the property accessed by a treacherous unpaved dirt road. The yard around the house was devoid of grass or weeds and swept clean for fear of snakes. Many farmers in the area were tenant sharecroppers who depended more on the production of "moonshine" whiskey for a living than on the sale of their share of the cotton and corn produce. Tenants came and went annually, usually by eviction, for failure to produce sufficient crops and operating an illegal "whiskey still" further reduced the landlord's share of the corn crop.

My brother Needham describes an event that occurred one Sunday afternoon when Daddy decided "to check things out on Trinity Mountain" as follows: "You could see through the house from front to back. Amos was not visible. It was obvious he was not under the house with the chickens, either . . . My father told my oldest brother, 'Henry, go around back and see if he went out the rear of the house.' Henry yelled, 'Daddy, he's running out the back . . . down into the woods.' My father saw Amos at about the same moment and yelled in a bloodcurdling authoritative tirade: 'Stop, you goddamn thieving son of a bitch,' or something close to that! Amos kept on hightailing it down the hill and into the woods. My father raised his double barrel shotgun and fired . . . twice! I thought I had just seen something unspeakably terrible. At age four, death was not a clear concept to me, but I was terrified at what I had witnessed. My mother stood transfixed to her spot in the front yard, her face frozen in an expression of horror. I thought Amos would be lying in the leaves and grass with blood all over him, like the quail from my father's hunting jacket that I had seen many times. He wasn't. As I was to learn later, the shots had been intentionally fired over his head, but that was not known to me at the moment." This and other events help describe Daddy's temperament. I do believe that he is in a good place now. I loved, respected, and feared him then, and I love and respect him now.

3

FARM LIFE (1938-45)

Plans for moving to the country were fraught with good and bad possibilities for me. I liked the idea of a chance to start over in an environment where I was not known as the timid, fearful soul that I was. On the other hand, my low self-esteem insured continued high anxiety in anticipation of a change of schools and a new environment.

The Great Depression eventually prompted our parents to buy a seventy-seven-acre farm, half in rock and cedar trees and half in cultivation and pasture, with an old dilapidated house and log barn. Construction of the house had begun in the mid-nineteenth century as a two-story log cabin with one room upstairs and one room downstairs. Having been added to over the years, by the time they bought it, there were six rooms plus a hall through the middle with large front and back porches, a tin roof over the dining room and kitchen, and a seventy-foot-deep well on the back porch. There was a smokehouse and chicken house in the backyard, a storm cellar on one side, and an outhouse toilet fifty yards from the back porch on the other side of the cattle runway. The first sheriff of the county had lived there at one time, and I have heard that he occasionally hung condemned men from a large limb of an oak tree that was still flourishing in the front yard when we moved there. In spite of these primitive circumstances, accessible only via an unpaved road resembling a goat trail and difficult to navigate in wet weather, so much land with house, barn, and fences for only $2,000 was a good bargain even in 1939.

Our father hoped we would learn to work with our heads but was determined that we would learn to work with our hands. And so we did. Year-round, the time to rise and shine was 5:00 AM that was still night during winter months. Bedtime was 9:00-10:00 PM unless homework took longer. We learned to milk cows, grow food for ourselves and our animals, and produce a precious little bit of cotton for market. When our first cow arrived at the farm, Mother was first to grab the bucket and learn to milk the cow. Daddy was milking before long, although he was at least sixty years old by then and had never before milked a cow. Soon we boys were milking also. For those who weren't milking, there were mules, hogs, and chickens to feed and wood to bring in. When Bermuda grass threatened to take the cotton one summer, Mother and her three sons, with their hoes and straw hats, joined with other field hands to hoe cotton until little grass or cotton either was left. In summer, there was always grass to cut and gardening and fieldwork to do. Plowing, planting, and harvesting corn, cotton, hay, potatoes, peanuts, and sweet sorghum were familiar activities for me from age eleven to sixteen when I went away to college.

The physically hard and often lonely hours and days of farm life without electricity or telephone, with few visitors or opportunities to visit, and isolation with little transportation was more difficult for some than it was for me. I welcomed the opportunity to be away from threatening human interrelations. I became interested in many things I had never noticed or experienced before. The old beehive was a curiosity and one to be wary of. Finding an occasionally edible peach or pear in a dilapidated and long-untended orchard was at first a satisfying discovery, and climbing the old pear tree to retrieve a fruit made it even more so. Collecting eggs and seeing hens incubate them and care for their young were new and interesting experiences as was the birth of a new calf. Encounters with reptiles were more frequent than ever before. And what could be more fascinating or entertaining than watching Budge or Toddy, our pet cats, dispatch a small garter, king, or hognose snake, which I had caught and kept for the occasion, during the noon-hour break from fieldwork? The hired hands also enjoyed these bouts in which the poor snake always lost.

One weekend when Daddy was home from the office, Mother became quite ill in the morning. She began to suffer unfamiliar pains and, within an hour, was having difficulty breathing. Daddy put her to bed in the downstairs log room, which was a bedroom at that time, because Grandfather and Grandmother Liston occupied another room across the hall, that would later be my parents' bedroom. Mother had collected eggs from the henhouse that morning, and we recalled having seen black widow spiders nesting there. Upon inspection of the dress she had worn earlier, Daddy found the crushed body of a black spider. He quickly stoked up the fire in the kitchen stove, started heating water in a kettle, and soon had her drinking water as hot as she could stand and as fast as she could swallow. She felt that her throat was closing, and he reasoned that heat should reduce the swelling. The treatment was not based on any prescribed medical treatment for black widow bite that I know of, but Mother was up and working again before sunset.

Times were hard, and farm labor was paid seventy-five cents per day until about 1940. Les Adkins and Willie Kennedy were black hired hands who worked with us for several years during the late thirties and early forties or until wages began to exceed one dollar per day. There were others, but I remember Les and Willie most vividly. When Daddy got home from the office each day, he would change clothes, help with chores, and walk around to check on things and see what had been done. The farmhands were always running short of money for tobacco by the middle of the week. So he would lend workers five, ten, or twenty cents as needed, keeping a record of the loan in his pocket notebook. When cash wages were paid on Friday afternoon, each worker's debt was subtracted from his pay based upon the notebook record. Sometimes a worker might have spent 20 percent or more of his pay for tobacco.

Not being a farmer himself nor having had much farm experience, and practicing law for a living, Daddy was not prone to spend a great amount of money on animals or equipment when the return on investment was so uncertain. Old mules were definitely less expensive than young ones. Big Ada, Little Ada, and Old Red are the best remembered of half a dozen elderly mules bought at Saturday auctions for rock-bottom prices over a period of as many years. Upon the arrival of a new mule almost annually, the farmhands were always anxious to count and examine its teeth and determine how much shoe polish covered the gray hairs on the animal's head. Daddy occasionally overheard criticism, most of which was behind his back, regarding his judgment exercised in the purchase of work animals. To counteract this criticism, on one occasion, he surprised everyone by having a young unbroken western mustang mare delivered to the farm. We named her Maud.

One Saturday morning, several young girl students from one of Mother's classes at Decatur High School came to visit at Mother's invitation and to see what farm life was like. After seeing the house, the girls were entertained by a guided tour of the barnyard where they were introduced to the pigs, Delphinium, Nasturtium, and Lakeside Sue, all feasting on kitchen slop at the trough. I felt shy and a little embarrassed for these city girls to see our circumstances. Since they didn't come at milking time, we boys couldn't demonstrate those skills. But we could and did take them to the pond where the cows and mules drank daily. We pointed out each animal by name as they grazed: the cows Emily, Betsy, and Bossy; the mules Big Ada, Little Ada, and Red; and the partially trained mustang mare named Maud. Feeling that I had failed completely to entertain the ladies in conversation and seeing Maud grazing nearby in the back pasture, I determined to show the girls a little horsemanship. By cautiously approaching and talking gently to Maud, I was permitted to pet her on the neck. In another second, I had a handful of her mane and was on her back with my feet pressed beneath her flanks. She took off like a jet-propelled rocket, literally and audibly expelling gas while accelerating for about fifty yards before suddenly twisting, turning, and standing on her front legs. I sailed through the air landing on my rear amazingly without any broken bones. Everyone was entertained, and I felt a sense of accomplishment in the episode.

Social interactions of any sort for us boys were not common. However, I recall a one-time event in which we brothers were visited by several boys from a couple of nearby farms. The neighborhood boys were out to meet their new neighbors. There wasn't much conversation as six or seven boys, ranging in age from perhaps eight to fourteen, wandered slowly about the barn lot. One would comment briefly on some common sight that would be followed by a grunt of acknowledgment from another and then a period of silence. In this process, we inspected the hog lot, the old hay rake, and the familiar old log barn looking in each stall and then the grain bin, before climbing the ladder to the hayloft. The loft was only about half-filled with hay at the time. So there was ample space on the floor of wide rough boards to sit in a small circle like Indians at a ceremonial powwow. The oldest boy present lived beyond two fields and a drainage ditch from us and was nicknamed Funny. Funny's talk soon focused upon girls and sex. He appeared to be the only one with any significant experience on the subject, to which his little brother also enthusiastically affirmed as he had "seen it happen." Funny removed his penis from his trousers and began to fondle it for all to admire its size. A few moments later, he was masturbating and urging everyone to join the contest to see who could ejaculate first. At age twelve, I was no stranger to masturbation, but never before under these circumstances. Not to be outdone, two or three other boys and I began to follow suit. The youngest boys simply watched in amazement or embarrassment or perhaps both. Funny won the contest. I was embarrassed that I could become no more than half erect, although I had never before knowingly performed with an audience. There was some debate as to whether another boy had actually completed the course or not. Consensus was that it was only urine or sweat but not semen.

On one of Daddy's inspections of the young cotton crop, he noted that numerous seedling plants had been clipped by a sharp instrument about a half inch above the soil surface on several rows near the road and the edge of the field. As a couple of neighbors had complained when our cows, on more than one occasion, had broken through the old fences and gotten into their fields, he wondered if someone in their group was paying us back or if some vagrant with a score to settle was indulging in a little mischief. He watched closely for several days and noticed that freshly cut plants were found daily in early morning. He concluded that the mischief was occurring after dark and started watching each evening for several hours until well after bedtime. His vigilance revealed no culprit, but newly cut plants continued to appear each morning. In desperation, he stationed us boys at sunset on a blanket in the woods across the road from the cotton patch with a double-barrel 12 gauge shotgun loaded with number 8 bird shot and with orders to shoot anything or anyone leaving the road and going into the cotton patch. He assured us that at the distance involved, we could not possibly harm anyone but might frighten them enough to teach them a good lesson. He would relieve us by ten o'clock and watch himself until the wee hours of morning. After a couple more days, the culprit was still at large, and fresh damage continued to appear daily. Upon asking Les for his opinion of the matter, we learned that rabbits sometimes do that sort of thing. Some years later, I learned that cutworms also may be the cause of such damage.

Under the guiding hands of knowledgeable instructors (Les Atkins and Willie Kennedy, black hired hands), I learned to harness mules and hitch to wagon and plow. As I was the oldest of three boys, I was usually first, but not last, to benefit from such instruction. Turning a weed patch into a plowed field ready for planting gave me a feeling of worthwhile accomplishment as well as a sense of superiority and control over this animal much bigger and stronger than I, although this required a day or more walking under a hot sun with one foot in the furrow and one on the ridge behind a mule, which amazingly could turn its colon wrong side out while defecating without missing a step. Many years later, I would wonder if one foot on the ridge and one in the furrow could have caused my left leg to be slightly shorter than the right. However, I felt great sympathy for the poor mules, particularly so when each in turn died of age-related problems and were dragged into the woods where they would be reduced to skeletons by hungry buzzards.

After cotton and corn were planted and rows established, this ridge and furrow walking was no longer required. Now it was time to plow the middles with a winged sweep plow for weed control and to throw a little dirt to the plants to support them as they grew. Here the need was to move at proper speed; too slow would not move enough soil to the small plants, and too fast would cover them entirely. Mules had better personalities for this sort of work than horses in general and Maud in particular. They plod along hour after hour without objection, except that by late afternoon, they begin to thirst for water and hunger for food and rest at the barn. Though they are not oriented at all times to see the barn, they know where it is. Thus, when the cotton was small and with rows running toward and away from the barn, the mules increasingly needed be held in rein when going toward the barn and urged on when headed the other way.

By comparison, poor, untrained, unbroken, young, energetic, and nervous Maud would simply go crazy at times. When harnessed to a cotton wagon, she stood on her hind legs and came down with her front half on the mule's side of the wagon tongue. After several severe beatings, administered by Les, Maud would start nervously when hitched to a plow. One morning, I took the plow handles behind Maud, with Les standing by. Things went well for about thirty feet down the row until the plow bucked slightly when a trumpet vine root was cut. Maud farted loudly and took off diagonally across the field with the plow hitting the ground every ten to twelve feet. I was fearful that the plow point would be buried at any moment in her rectum. She finally stopped at the fence by the edge of the field, wild-eyed and blowing hard. Les cut a large limb from a persimmon tree and hooked the animal's traces to the tree trunk before beating her mercilessly until there were large lumps on both of her sides accompanied by bleeding. Maud was sold at auction a short time later. Poor Maud was always in trouble for just being herself and responding to her natural instincts. I often wondered if the discipline imposed on her might have been more harmful than helpful. But who was I to argue with one of Les's experience?

My brother Rob and I, accompanied by Les, were cutting firewood with the crosscut saw one Saturday morning in late fall. Les and I were tugging at the saw as Rob attempted to split a chunk of wood with the axe. Suddenly, the axe struck a glancing blow and bounced deeply into his ankle. Blood gushed out immediately in spurts with each beat of his heart. I suppose we were a quarter of a mile from the house with a mule-drawn wagon. I was paralyzed with fear that my brother was bleeding to death before my very eyes. It had rained earlier, and the ground was wet. Les calmly grabbed a handful of wet soil with one hand while manipulating Rob's ankle to hold the cut widely open with the other before pushing the wet sand and clay deeply into the wound. He then lifted Rob onto the floorboards of the wagon and drove the mules at a fast gallop with the three of us bouncing in the wagon all the way to the house. As electricity and phone service had not yet reached so far out in the country, we were fortunate that this happened on the weekend while Daddy and the car were home from the office. Rob made it to the hospital in time and survived to become a very skillful thoracic and general surgeon years later.

Hauling cotton in a mule-drawn wagon six miles to town and the gin was an exciting first experience. I sat high on the cotton at the front of the wagon beside Les watching his every move and listening to his every command while marveling at his skill and at the ease with which he controlled the mules. Navigating several blocks of paved streets in auto traffic seemed even more challenging, but it was just all in a day's work for Les. I had already learned that mules easily defecate without pausing while pulling a plow. So I wasn't too surprised, although a little embarrassed, to see them dumping without hesitation on the streets of Decatur and in the presence of so many witnesses. On arriving at the gin, we fell into line with other mule teams and wagons awaiting our turn to have the loaded wagon weighed and then emptied by a maneuverable suction tube before reweighing the empty wagon and getting a receipt for our cotton from the office manager. The return home was not nearly so comfortable. With no springs and with wagon wheels of wood and steel, every bump or piece of gravel in the road was felt until I wondered how much more jarring and shaking my teeth and other organs could tolerate.

When I was about twelve or thirteen, Mother made white shirts and pants for each of us boys from empty Sunnyfield flour sacks, which she boiled in a wash pot to remove as much of the colored print as possible. However, some of it still showed, which was a little embarrassing but not enough to prevent our walking to town at noon on Saturdays to see as many movies as we could before dark. This was reward that Mother felt guilty for permitting but that Daddy agreed to in return for our hoeing cotton or finishing some other chore that the paid hands had not completed on Friday. The three of us would start together walking two miles to Austinville on the rough, unpaved road and another four miles to Decatur first on gravel and then hot asphalt. By the time we reached Second Avenue, I would be a hundred yards ahead of Rob and he a similar distance ahead of Needham. Such were the benefits and penalties of age. We each had less than a dollar in change that would be enough, if we could get there in time, to attend the Strand and Princess theaters on Second Avenue and the Roxy Theatre on Bank Street before catching a bus back to Austinville and walking the rest of the way home. This relaxation of rules—permitting us to see such classics as the Lone Ranger, Tom Mix, and the Three Stooges—was appreciated and enjoyed beyond all imagination. I would laugh at the Three Stooges until I could barely breathe. Also, there was more to enjoy than the movie itself. Simply entering the darkened theater was an exciting though temporary escape from reality to which I often dreaded to return.

I must have been of age thirteen or fourteen when we decided that the old log barn was no longer safe or adequate for our needs. Brother Rob and I assisted Les, the master carpenter, in building a new barn of raw lumber with a new tin roof located near the pond where all the animals went to water. I suppose that we were of some help driving nails, but Les provided the brains and most of the brawn for the operation. It was much nicer now to throw hay into a loft that didn't leak and corn into a tighter and larger bin; also, there were more and better stalls for the animals. Mother did a watercolor painting of this new barn with the old pear tree in the foreground that is framed and now hangs on the wall of my study as I write.

Les was confident, calm, cool, and somehow aloof. He was always in control and had most of the answers for dealing with problems in the world in which I was living. Les usually had a little smile, suggesting that he knew something no one else did while resting under a tree sharpening his knife or whittling on a stick. I was sure that he did know a great deal more than most of us about many things. The other workers always respected him, and I don't think that even Willie—who was younger, bigger, and stronger looking—would have wanted to tangle with Les. On one occasion, Daddy bailed Les out of jail after he had been arrested for fighting somewhere with someone over the weekend. Les was apparently unhurt but didn't talk about it. I was shocked and grieved several years later, after my three years of military service and while in college, to learn that Les had been found dead, apparently murdered, beside the railroad tracks between Cedar Lake and Decatur. Although I had not seen him for years, I felt the emotional blow of having lost a good friend.

Old Red was probably the least expensive mule that Daddy ever bought. I believe he lasted two summers before we found him dead in his stall one morning. The old mules often died on cold winter nights with food in their mouths, but Old Red left in late summer. He couldn't wait. We decided to bury him in the pasture rather than drag him to the usual place in the woods. So we boys began one morning to dig a grave with pick and shovel. It turned out to be a much bigger job than anyone had anticipated. By noon, we realized that the hole was only a fraction of what would be required. Also, legs extended in rigor mortis further complicated the problem. To solve the unfortunate situation, his limbs were broken with the axe and folded against his body that was placed catty-cornered in a shallow rectangular grave. He was covered with six to twelve inches of soil and partially recovered several times thereafter until all scavengers lost interest. Years later and after the farm had been sold and subdivided, I pointed out to someone, while riding through to reminisce, that we had buried a mule in that lady's front yard.

For several years before electric power arrived, a typical winter evening at our house, after chores were completed and the table cleared of supper dishes, would find the entire family in the dining room warmed by a roaring fire in a wood-burning heater. Mother would be seated at a table in one corner, which was lit by a small kerosene lamp, grading test papers or preparing for tomorrow's classes at Decatur High School. We three boys would be seated, one on each side and one at an end of the dining table—which was lit by a large kerosene lamp in the center. This table was once Grandmother Liston's dining room table and now is in Jan's home in West Monroe, Louisiana. Daddy would alternate between sitting at the other end of the table and walking around it to be sure we were doing our homework. I remember crying in frustration one evening because I just could not understand my algebra assignment. I will never forget Daddy's attempt to encourage me by saying that he had encountered many men in his life who were much smarter than he, but he had never seen the color of the man's eye that he could not beat by outworking him. That thought has stuck with me ever since.

While Daddy was a strict and sometimes harsh disciplinarian, he did like to cool off at times on a hot summer afternoon. Occasionally, he would invite his sons to get in the car and go swimming. Several miles away, there was a swimming hole where Flint Creek passed close to the Somerville Road. Someone had erected on the creek bank a makeshift diving board from a cypress plank, about twelve inches wide by three inches thick, which could spring you pretty well if you weighed enough. We boys were relative lightweights at that time. But Daddy, in his midsixties with a little bulge at the waist, could spring high off the board for a nice dive and sometimes a splashing belly buster. Fun was had by all on these occasions during which we kept the water sufficiently agitated to keep the reptiles at bay. But we were always aware of their presence by their heads sticking out of the water a short distance away.

Electricity finally reached our farm in the early 1940s. We three brothers had shared the upstairs bedroom in the original log cabin portion of the house until now. There were two beds in that room so that only one could enjoy a bed to himself. Since I was oldest and strongest at that time, I often took what I wanted if I could get away with it. However, no one wanted to sleep alone in the cold of winter when the wind was felt coming through the walls and snow from a window blown open during the night had dusted our blankets. "Again, if two lie together, then they have heat; but how can one be warm alone (Ecclesiastes 4:11)?" The sharing of body heat with several bodies together was quite beneficial. However, the room became more crowded as we grew. Being the oldest and largest at that time, I benefited most during winter from a smaller body surface to volume ratio, which helped conserve body heat. At any rate and for whatever reason, I was permitted to convert a downstairs room into my own bedroom that opened on to the back porch and that had served for storing potatoes. I saved for and bought a small radio about the size of two bricks, which was placed on the windowsill beside my bed, and began to enjoy nightly the wailing sounds of old-time country music ("Sleeping at the Foot of the Bed," "Hot Cabbage in a Little Tin Cup," and "I'm Moving On") while looking through the screen at the stars until sleep overtook me.

Stella and Leonard Burns were resident farm laborers who worked for us as little as possible. They lived across the road and down the lane through the woods in a two-room log cabin with a large stone fireplace in the living-bedroom and a cast-iron wood-burning stove in the kitchen-dining room. Farm wages had increased from seventy-five cents to a dollar a day, and Leonard and Stella had free rent in addition to wages. They had many excuses for not working regularly, but they managed to take hayrides to Nashville to the Grand Ole Opry. I enjoyed visiting occasionally to hear about their travels and to see their gum-chewing puppy that slept in the warm ashes near the backlogs in the fireplace. They were replaced eventually by Gene Slate and his wife who promised to be better workers. However, appearances could be misleading as Gene shortly told Daddy: "To tell you the truth, Mr. Long, I don't like nothin' about a farm!"

When I was thirteen or fourteen years old, my brothers and I took advantage of what then appeared as a great opportunity to make some money selling Christmas trees. After all, we had forty acres of rocks and cedar that even a few cows and goats couldn't convert to anything of much value at the time. With saw and axe, we cut and loaded a mule-drawn wagon full of cedar trees. The next day, we drove the mules and wagon six miles to town and peddled Christmas trees from door to door on streets in west Decatur for the price of twenty-five to fifty cents each. We must have made close to five dollars that day. I don't remember how the money was divided among us. I hope that I was forced to be more charitable and fair about it than I naturally was inclined to be at the time, particularly with smaller and younger brothers.

There was an elderly black man, whose name I don't remember, who came around annually in the fall with his mule, a mill for crushing sweet sorghum, and copper vats in which juice from the mill was cooked over a wood fire. As the juice boiled, it was raked slowly with a wood paddle from one compartment to another and from one end of the vat to the other while it slowly changed from raw juice to syrup and then to molasses. When ready, the molasses was drained into gallon-size glass jugs and corked. We boys, with additional help, cut the sorghum by hand and delivered it by wagon from a small patch across the road to the old man's mill, set up in our front yard, where his mule walked in a circle all day turning the mill to crush the sorghum. The job was completed in a couple of days and the molasses divided equally between the old man for his service and us for consumption with hot biscuits, corn bread, and pancakes.

Another thing about farm life in those days that must be mentioned is the Saturday night bath. Of course, we usually bathed more than once a week; however, Saturday night was a "must" in preparation for church on Sunday. In cold weather, water was heated in a kettle on the wood-burning cookstove in the kitchen and poured into a wash pan with water drawn from the well to obtain the desired temperature. Then the actual sponge bath might be performed in the kitchen or the bedroom, depending on the need for privacy by the individual bathing. This was standard procedure until Christmas of 1946 when, with army pay and as a Christmas gift to the family, I ordered a bathtub from Sears & Roebuck to be placed on the back porch. It was rigged with a garden hose to drain it into the yard, but water still had to be drawn by hand from the well and heated on the stove to enjoy this luxurious tub bath. Shortly thereafter, a wall was built to enclose the new tub, and later, an electric pump was put on the well, a septic tank installed in the yard, and the dream of plumbing with a complete bath and indoor toilet became reality. By that time, I had gone and would live there again for only a year after discharge from military service.

It was in the seventh grade (1939-40) that the size of my feet became quite noticeable, and it was there that I acquired the nickname "Feets." How I hated that name and dreaded being greeted by it! Nevertheless, it caught on quickly and was the name by which I was best known for the next three years. Every time I heard it, I wanted to hide in a hole or just be a fly on the wall. But no such privilege would be granted. I hated the name and began to hate all who spoke it. I would have liked to seriously harm them but didn't have the nerve to satisfy my fantasies. Instead I learned to avoid eye contact and to walk with stooped posture and head down. I intentionally avoided encounters with certain people when possible. I sympathized greatly with a poor, ragged, barefoot, and always snotty-nosed boy laughingly called "Button Eyes" by many who rode our school bus. I suspect that his bulging orbs must have derived from an abnormal medical condition about which he could do nothing.

During the 1943-44 school year, when I had turned fifteen, I rode bicycle for Western Union delivering telegrams all over the city of Decatur, working after school in the afternoons until 5:00 PM, when Daddy closed his office and headed to the farm, and on Saturday mornings. Mr. Couch was a short thin man with protruding teeth and pockmarked face. His employees all had time cards that were stamped upon arrival and again when checking out for the day. The idea of such fairness in pay regulated so precisely by the clock with each act of stamping in, stamping out, and seeing my hours accumulate as I returned my card daily to the rack on the wall was a source of considerable satisfaction. Messages had to be delivered in all kinds of weather. In stormy weather, one wore a raincoat and hat but often got pretty wet anyhow. If lightning wasn't too bad, you just took your chances. I don't remember what the hourly pay rate was, but it was the best money I had ever made. Some weeks, I made almost ten dollars. Mr. Couch thought the world of me and of my father, whose office was across the street above Kuhn's five- and ten-cent store.

In the summer of 1944, my brother Rob and I were allowed to work part-time away from the farm for a few days each week with the Decatur Street Department for extra spending money. This was hard and hot labor paving parking lots and mending street potholes with asphalt. Our overseer was Tom Fred Nelson, under whom we and several young black men shoveled, raked, and tamped hot asphalt. I was fifteen years old and had a driver's license that entitled me to drive the dump truck at times and spread the asphalt as well as walk in it, shovel it, rake it, and tamp it. This did not sit well with Rob, aged thirteen, who believed that he could drive as well as I. And as unfair as it seemed to my brother, I took great satisfaction in driving the truck, backing up, and dumping gravel or asphalt while the rest of the group grunted and shoveled. We all sweated and all earned a few dollars a day by then; I don't remember exactly how much. But one dollar was still a lot of money in those days. I believe that Cokes were still a nickel or no more than a dime.

During my last year of high school, 1944-45, I worked part-time after school and on Saturdays at the A&P grocery store, a block from Daddy's office on Bank Street, until time for the office to close and for us to head home. In those days, duties as an A&P clerk were not strictly categorized. I started sweeping floors, then stocking shelves, and, shortly thereafter, helping out at the cash register during rush periods, or doing anything else that might be needed. Learning to quickly make change without using pencil, paper, or adding machine was an important accomplishment that I thought must surely put me ahead of many folks, at least in that respect. However, my pride was badly wounded one day when a man of questionable character made a purchase while I was working the register. He had a few items that totaled let's say seven dollars and seventy-two cents (I don't remember the exact amount). He gave me a paper bill, which I placed in the proper register drawer, before counting his change back to him: "seventy-three, seventy-four, seventy-five cents, eight, nine, and ten dollars, sir, and thank you very much." He immediately wanted to know where his other ten dollars were. I was thunderstruck. I was certain he had given me a ten-dollar bill, but he claimed it was a twenty "with Jackson's face as plain as day on it." Mr. Roberts, the store manager, quickly intervened and, after some discussion, gave the man another ten dollars. He then explained to me that while I probably was correct, I had lost my proof by prematurely putting the man's ten-dollar bill in the register when I should have placed it beside the register on the counter until the making of change was completed. An important lesson I still remember.

4

PARENTAL AND OTHER INFLUENCES

Mother and Daddy seldom entertained but frequently visited back and forth with a few close friends. I do not recall their ever giving parties or attending social gatherings of more than two or three couples, except for church-related functions. Conversation with friends was often animated and jocular but never related to sex that was studiously avoided and never joked about. Conversation with children in adult company was rare as we were "to be seen and not heard." Displays of affection between parents or between parents and children such as hugging, caressing, and kissing for more than a peck generally were not seen in our home. I often heard Mother say "I love you" after we children were grown and particularly in her later years, but such expressions were not frequent during our childhood years.

Mother worked hard and continuously at everything she did. She believed that sitting idly with nothing to do was not only a waste of time but also sinful. "An idle mind is the devil's workshop," she often said. And soon thereafter, if we did not become gainfully occupied, we were directed to find and carry stones to a retaining wall separating our front yard from the unpaved road in front of our house on the farm. This wall was in process of construction for several years. There is no doubt that she loved her children dearly and would have made any possible sacrifice for them. However, I have wondered if she ever found time to simply play with her babies as many mothers often do, tickling, playfully pinching, laughing, hugging, etc.

As a child, I longed to play sports with a passion but never played or even tried out for any team sport in the grade, junior, or senior high school. I was too young from the coaches' standpoint and too busy with farm chores to justify the time and extra travel that would have been required from my parents' point of view. For several years on the farm, I had a rubber ball, half the size of a regulation basketball, which I dribbled frequently enough to keep the ground bare of grass halfway around a large oak tree, on the trunk of which was nailed a metal ring slightly larger than the ball. This was by no means a basketball goal, and there was no backboard, but it served that purpose for several years. Once in a while, my father would accompany me to a basketball game at Austinville High, a county school, where I attended seventh, eighth, and ninth grades. I always sat with him throughout the game and always returned home with him. I recognized many of the kids at Austinville High and knew many of their names. After all, I rode the school bus with some of them, but I had no close friends there.

In the country, our closest neighbors were a half mile or more away and were mostly sharecroppers who had farmed all their lives and with whom we had little in common, or so we were led to believe. In town, the distances to neighbors' homes had been much less, but always there was a supposed gulf of some sort—moral, intellectual, economic, or indefinable—which usually separated us from them and limited our contacts with others. In junior and senior high school, I was socially isolated and emotionally disturbed to the point of having trouble concentrating, thinking about or remembering facts and figures associated with academic studies. In the back of my mind was always the fantasy of getting even with someone for some inflicted wound, be it real or imagined, or fearfully planning how to avoid further emotional wounds. My brothers, who lived in the same general environment as I, were both consistent honor students from first grade through the University of Alabama Medical School. I would be a college senior before ever earning such grades at school.

To help overcome my extreme shyness, Mother decided when I was in eighth grade that I should enter the school's oratorical contest. The thought of such a thing made me want to run as far away as possible. Of course, that could not happen because it would not have been permitted. Instead I began to try to write a five-minute speech, and as strange as it may seem, I have no recollection today of either the subject or content of that speech. However, I vividly remember the sheer terror that gripped my whole being at the thought of standing onstage before the entire high school assembly to do anything, much less speak. But the day did come. And I did stand at the podium before the assembled crowd. I could not look up. I could not speak. My hands shook violently as I held my papers. I was in panic. After a few agonizing moments and without performing at all, I was thanked for my efforts and led offstage by the teacher in charge. Daddy and Mother had both attended and were now in the front seats of the car where I watched and listened from the backseat as we rode home to hear Mother's comment: "I will never try that again!"

Of course, something similar had been tried earlier in second grade before we moved to the country. Mother had required me to study violin for a while at which she was quite competent. She accompanied me at the piano on this occasion as I serenaded the ladies of the Lafayette Street School PTA to the tune of "London Bridge Is Falling Down." I was much shaken by the experience but managed to get through it without actually falling down myself. At this time, my self-esteem was already low, but not nearly as low as it would be after six more years.

Mother was very upset by a radio broadcast of the now-popular Christmas song, "I Saw Mommy Kissing Santa Claus." She believed that this was detrimental to the moral fiber of the nation and that it threatened the sanctity of marriage in the minds of young people whose ideas on morality were still in a formative stage. I believe that she would have agreed with George W. and Laura Bush on many issues, but she would have been very unhappy to see and hear the First Lady recently name the above song as her favorite Christmas carol.

Undoubtedly, my best and most lasting friendships during grade school, junior and senior high school were with the Speake boys, Dan and Neal. I never saw their mother, who died when they were quite young. But she and my mother had been close friends for years and attended the same church. Dan was in my grade at school, though almost two years older, while Neal was my age but a grade behind me. They lived in a large two-story house on Ferry Street in Decatur with their father, a mechanical engineer and widower till his death. During the years from the time we moved to the farm in 1939 through my middle school days, it was not uncommon for the Speake boys to visit on Sunday afternoon at the farm and sometimes even spend Saturday night for an extended weekend as well. Likewise I would spend the night with the Speakes in Decatur. Dan and Neal were big, muscular, athletic, and good-looking boys. In high school, they both played varsity football, and Neal later was a defensive end at Sewanee—the University of the South.

The fact that Dan and Neal occasionally returned home with us from church for a Sunday afternoon visit had no effect on the customary sequence of family activities for the remainder of that holy day. After lunch (dinner in those days), there was always a mandatory period of hymn singing, with enough hymnals to go around, during which all verses of four hymns were sung. Then there was reading of scripture followed by half an hour or more of scripture memorization. This period could be ended only through satisfactory individual recitation of newly memorized verses. These rules applied to family members and visitors alike. In spite of such hospitality, Dan and Neal never tired of visiting the farm, playing rough-and-tumble hayfield football in the late afternoon, or reenacting with wood swords and quarterstaffs the capture of a royal caravan by Robin Hood and Little John in the woods across the road, or catching snakes, hunting rabbits, etc. until their father appeared near dusk for a short visit before taking them home.

Late one Sunday afternoon in December 1941, Mr. Speake drove into the yard for a short visit and to retrieve his boys who had enjoyed a long weekend on the farm with the Long boys. It was December 7, to be exact. Electricity and radio had not yet reached our farm. Our visitor's first words were "Have you heard the news? The Japs have bombed Pearl Harbor!" As he described what he had heard about the loss of over two thousand American lives and the sinking of numerous US Navy ships that very day, some of which were still burning by all accounts, the enormity of the disaster slowly began to dawn on us all. I felt an anger welling up inside as strong as or stronger than what I experienced when being teased at school. The following day, I wrote a letter to President Roosevelt volunteering my services to the nation and asking for his suggestions as to how thirteen-year-old boys might serve in the national defense. Answer to my letter arrived from the White House eight weeks later. In it Mrs. Roosevelt, through her secretary Edith Helm, expressed her gladness to know of my patriotic enthusiasm and desire to help my country. However, she felt that a group, such as I wanted to organize, should be a purely local one at this time and that I should talk with an older person in my community about this idea. I was happy to finally receive an answer but frustrated to realize that I was yet too young to do much about it.

Dan and Neal shared a bedroom in the upstairs of their house into which I was told that Dan would sometimes bring in a drive-in waitress after-hours. From what I could understand, Neal's earliest sex education came from these live demonstrations. After enjoying the advantage of such training, Neal must also have become more adept at such matters I believed. Our friendships remained firm but became increasingly strained by my lack of such basic training and experience. I remember a later time when Neal talked me into attending with him a junior-senior high school dance at the historic old bank building in Decatur. Neil was trying to be my buddy and simultaneously entertain several girls while I nervously longed to be back on the farm. It wasn't that I didn't earnestly long for female companionship and camaraderie. It was simply that I felt as awkward and nervous around girls as I would have been at another speaker's podium.

I transferred for tenth grade from the Austinville county school back to the city school system at Decatur High School where Mother was now teaching art, history, and Latin. My popularity as "Feets" was not widespread here, although the size of my feet did not go unnoticed. But somehow and from somewhere, I became known as "Lucy Long," a name that stuck with me through graduation from high school. I despised it with a passion, not only because it was another nickname and a put-down but because it carried a humorous (for some) feminine connotation that I highly resented. Still I lacked the courage and the encouragement to stop it. Today, I would advise any such child as I was then, regardless of size or age, to verbally demand that the tormentor cease and desist, and if compliance was not immediate, to plant a fist squarely in someone's face, regardless of consequences. I believe that I did seek to foster that attitude in my own children as circumstances seemed to require. Unfortunately, in such matters, Daddy's opinions did not often take precedence over Mother's. Mother stood for peace without violence or confrontation. However, my youngest brother, Needham, states that he overheard Daddy remark to Mother on one occasion, "Sara, why don't you leave these boys alone? You'll make them so that any woman who knows the family won't want to marry them."

To say that my social experiences from birth through high school were limited would be a gross understatement. I graduated from high school four months before my seventeenth birthday having never danced with a girl and having had only one date, more or less arranged by the mothers, with the daughter of a schoolteacher friend of Mother's. Betty Hunter was a very pretty and popular, although shy, junior in my senior year. Her mother met me at the front door. With sweaty hands, I gave Betty a little bouquet, homemade with Mother's help, and we left in Daddy's car for the Roxy Theatre. I do not recall the movie, nor do I recall holding her hand or even putting my arm across the back of her seat. Afterward, we went to a drive-in for milk shakes that were imbibed with scant and labored conversation. By then, it was time to take her home where her mother waited in the living room. I saw her to the door and, with great relief, got back in the car and drove home.

In grades one through twelve, I often would have preferred being a fly on the wall than a student at school. The overall effects of many of the puritanical principles by which we were raised and the situations that we encountered from birth through high school greatly reduced our opportunities for social interaction and for the normal development of social skills. They also undermined self-esteem and created a sense of uneasiness toward the opposite sex. Certainly, our parents could never have imagined the negative results that would follow their dedicated efforts and good intentions. However, in spite of such negative influences, all three brothers were thoroughly imbued with a determination to persevere always and to seek to be the best at whatever they did. The results of these qualities eventually surfaced in one way or another in each of their lives. Needham would become a pathologist with triple specialty board certifications, heading his own laboratory and serving hospitals in a hundred-mile radius around Anderson, South Carolina. Robert would become a skilled heart surgeon with publications documenting his contributions to science and, later also, a general surgeon practicing in Florence, Alabama. Henry would become a college professor and agricultural consultant, would serve assignments to Egypt and Brazil for the United Nations, and would be elected to the Louisiana Agriculture Hall of Fame for his services to the betterment of Louisiana agriculture.

5

WASTED YEAR AT COLLEGE (1945-46)

One of the best days of my life, to this point, occurred May 28, 1945, on which date I received a diploma certifying that I had satisfactorily completed a course of study prescribed for graduation from Decatur Senior High School. In three more months and twenty days, I would be seventeen, old enough to join the army with parents' permission. More than anything else in life, I wanted combat training with all the toughness that I could possibly acquire by it. I wanted to be so dangerous that no one ever would dare abuse me in any way. Problem was that parents would not consent.

As an infantry officer in World War I, Daddy did not want me exposed, at my tender age, to an environment that he and Mother both felt would not be conducive to my continued proper development. All my pleading was wasted effort. I threatened to run away and lie about my age but was assured that I would not succeed at that either. They wanted me to go to college for at least a year and continue my education. From my point of view at the time, I had learned nothing useful from eleven years of education (we all skipped a grade in elementary school). So how could I possibly benefit from one more year? I was fed up with books, teachers, and particularly with unfriendly older fellow students with whom I had nothing in common.

Mother's only brother, Robert Todd Lapsley Liston, two years her elder and of whom she was intensely proud, was then president of King College in Bristol, Tennessee. This Presbyterian school was proud of being "austerely, determinedly small, determinedly Christian, determinedly scholarly." Mother had great faith in the belief that her brother would prove uniquely equipped to help her troubled son find the right path. After all, Uncle Robert's "real education came from the lips of our father and mother" and "the schools from which his formal training came included McCallie, Davidson, Union Seminary in Richmond, and the University of Edinburgh" in Scotland from which he earned his doctorate. With due respect, I could not at the time imagine getting any useful help from such sources. However, as there seemed to be no other possibilities, I was resigned to attend King College for a year on condition that my parents give official consent to my joining the army immediately thereafter. We all consented, and I departed after several days for Bristol to live with Uncle Robert and his family on campus and work at a summer job.

Uncle Robert found for me a summer job with Mr. Irby, who ran a sizeable nursery and greenhouse. Aunt Maria was a sweet, loving person and mother of two boys of her own, my cousins Bobby and Miller. Uncle Robert was always up early for his daily exercise walking through the campus while Aunt Maria prepared the family breakfast. After breakfast, I walked down the hill to the campus gate to wait for Mr. Irby to pick me up on his way to work in a battered old red pickup truck. One of our first jobs almost daily was cutting gladiolas, roses, and other flowers for delivery to hotels, restaurants, weddings, etc. Next, greenhouse benches were watered, plants were potted, others were rearranged, and there were always pots to move and cleaning up to do. Work was six days per week, ten hours per day. Payday was every other Saturday, and I paid my uncle regularly for room and board. Still I had spending money to spare, or so I thought, which seemed more than adequate at the time. I would begin to learn later how limited my tastes really were.

Most of the nursery workers were old folks in their late thirties at least. Working beside a thirty-eight to forty-five-year-old woman in the nursery was not much more threatening than working with another male. After all, she was old enough to be my mother. My discomfort and uneasiness were more pronounced with younger females than with males but were based more on difficulties with interpersonal relations in general than with sex in particular. I was never at ease in any social situation requiring interactive conversation. The larger the crowd, the more uncomfortable I invariably became.

Word spread on campus that a nephew of the president would be among the new freshmen boys in September. There was some interest at the girls' dormitory about this new prospect when students checked into their dorms for the fall semester. I was assigned a second-floor room in the boys' dorm with Henry Dendy, the wild, comical, and talkative son of a Presbyterian minister, and Ed Zink, a tall, lanky boy with more interest in auto engines and motorcycles than books. Not a single occupant of our room was a serious student at heart.

On Sunday evenings after church, it was customary for many of the boys to congregate in the first-floor sitting room of the girls' dorm and gather around the piano with the ladies present for singing, listening, conversing, or just socializing generally in small groups or in pairs. Sue was an attractive freshman girl who played piano well and often led the singing or just entertained with a little boogie or jazz piano until Mrs. Inez Morton, dean of women and in charge of the dorm, announced that it was time for the boys to leave and the girls to be in their rooms. Sue seemed especially friendly toward me in spite of the introverted social cripple that I was. For me, this was both flattering and frightening. Her aunt was the head college librarian. And it seemed that Uncle Robert was loved by the entire faculty and most of the students.

Uncle Robert presided daily at morning convocation, often preaching a minisermon as well. Bible was a required course each semester. I remember that the fall semester dealt with the parables of Jesus, an excellent subject taught by an aging and retired minister, who didn't bring much enthusiasm to the course that I could appreciate. I read scripture irregularly and was not regularly prayerful. I found little benefit from reading and praying at that time. I found excitement and some strange satisfaction on one occasion by squirting other similar-minded boys with fire extinguishers in the dormitory halls until the floors and walls were covered with foam. The guilty were soon identified and punished by the dean of men with special extra duties in addition to cleaning up the mess. On another occasion, my roommates and I tossed condoms filled with water off the roof of our dorm during morning break between classes, watching with glee as the exploding water bombs soaked students and some faculty below. Everyone suspected who the guilty culprits were, but there was no proof, and we admitted nothing.

King College had suspended its intercollegiate athletic programs since the beginning of the war because so many athletes had been taken by the draft. There was a nice basketball gym with rows of empty bleachers being used only for physical education classes. Ed Zink and I plus several other would-be athletes and a couple of returned war veterans, who had been athletes a few years earlier, implored Coach Young, who had coached basketball several years earlier, to organize a team and schedule games for the coming season. Coach did not want to do it, considering it a practical impossibility to floor a competitive team on such short notice and with such limited talent. However, we prevailed on him with some arm-twisting from the administration, I suspect. We played about sixteen games on a home-and-home schedule against Tusculum College, Carson Newman, Martin Methodist, Maryville College, and East Tennessee State, among others. We lost every game, but all eight or nine of us lettered, and all got lots of playing time. I remember that in one game, I was high-point man with six points. That was a great experience for me at the time, although in looking back, I suspect that Coach Young must have felt at least a little like a Special Olympics coach.

The editor of the college yearbook, with support perhaps from the librarian and the dean of women, prevailed upon Sue Buchanan and me to pose for yearbook pictures, as if out on a bonafide date to the movies, the malt shop, etc. We went on this date with another couple who were upperclassmen. I did not anticipate that we would have dinner at a real restaurant before the movies, and I barely had enough money to cover dinner expenses for my date and me. As a result, Sue had to bail me out at the theatre. The pictures taken were used in a full-page spread at the back of the 1945-46 yearbook to illustrate some of the beneficial and healthy social activities at the college. Sue wrote on that page in my yearbook: "Well, Henry, here we are more stupid looking than we really are. I've enjoyed this year with you, and any time you want to go to the show, I will get a ticket, and we'll both go. Best wishes to a quiet but real nice boy—Sincerely, Sue."

I don't remember that Sue and I ever went out on another real date. By the middle of the fall semester, she was dating an upperclassman pretty steadily. I certainly had no justifiable reason to feel jealous. But I was. I watched her several times from my dorm window being escorted from her dorm to her beau's car before they drove away for the evening. I actually harbored a secret grudge against her for the remainder of the school year, as if I had been betrayed. My roommates and I group dated a few times with other King College and Sullins College girls. These were all one-time meetings resulting in no special friendships. A year later, after too many drinks and while stationed at Clark Field in the Philippines, I wrote Sue a terribly ugly and profane letter, showing it to army buddies as just reward for a "Dear John" from her, which of course was a lie. They dared me to mail it, which of course I did. Within a short time, I began to feel terrible about what I had done. Years later, I tried unsuccessfully to find her address online through my computer to at least let her know what a sick human being I had been and how sorry I was for my action, which has now been a load on my conscience for almost sixty years.

By the end of the spring semester, it was apparent to anyone who knew the facts that Henry Long was not breaking any academic records. The dean of men called me to his office one day for a talk. He advised me that I obviously was not in the mood to do serious college work at this time, and he thought that it might be in the best interests of the school and me to take a break from college, at least from King College, and try something different. I had earned Ds in Bible, Ds in history, Cs in chemistry, Cs in math, and Bs in English. I assured him that I had exactly such plans in mind and that I would not be back next fall but would be in the army. We shook hands, and I never spoke with him again. Uncle Robert denied to Mother that such a conversation ever took place. Perhaps Uncle Robert didn't know.

In my opinion, unless I've missed something, my year at King College served no useful purpose and did waste time that would have been better spent in the military. Of course, I had enjoyed very much playing basketball; however, in retrospect, that also probably contributed little to my much-needed development and education.

6

MILITARY LIFE (1946-49)

When the time came for my parents to make good on their promise of official permission for me to join the army, Daddy balked at my signing up for combat infantry or even for enlisting in the navy as an alternative. He believed that the air force, which then was a branch of the army, would be best for me. Therefore, the *Decatur Daily* reported that on June 1, 1946,

> Judge W. H. Long of the Morgan County court witnessed his son, William Henry Long III, being accepted for enlistment in the army air corps . . . His son's enlistment . . . was 29 years to the month from the time the Judge himself enlisted in the United States Army infantry . . . Memories of his own 22 months' service in World War I, from which he emerged a major, must have been crowding through Judge Long's mind as he stepped down the hall from his own office to that of the recruiting service to watch his seventeen-year-old son begin an army career.

I could hardly wait for a change of scenery. Deep down, I still loved and respected my parents, but I was filled with all sorts of frustrations and harbored a lot of anger for reasons I would have had difficulty defining. The next day I traveled by bus to Montgomery, Alabama, and to Camp McClellan for processing and swearing in. My first night in an army barracks was an education in itself for me. I had never before heard so much profanity used by so many so continuously. I swore to myself and to God before going to sleep that night that, live or die, the first person to call me an s.o.b. would have to suffer, regardless of name, rank, or serial number.

Basic Training. My first duty assignment the following morning was cleaning latrines with a detail of approximately a dozen other inductees. When a six-foot slightly obese mountain boy addressed me saying, "Hey, you son of a bitch, throw me that rag over here." I asked if he were talking to me. He said that he was and repeated the request. I was now a tall and wiry six feet five inches although only weighing one hundred sixty-five pounds. I picked up the cloth and took three or four quick steps toward him before striking him full in the face with the rag held in my fist. Apparently, I had moved fast enough to surprise him. At any rate, I pivoted on my left foot and attempted unsuccessfully to break his left leg backward at the knee with a hard right kick. I did hurt him but not as much as I had hoped. When the NCO in charge entered the room a few seconds later, I was punching and pedaling backward to avoid a bear hug. We both were ordered down on the floor for ten push-ups, nose to the wet floor, and we were given extra duty the next day working together during all breaks and into the evening. We worked together without further problems, and I never saw him again afterward.

The second day after arriving at Camp McClellan, we were given shots, vaccinations, medical inspections, IQ tests, and much of our GI clothing issue. I received most of what was needed with the notable exception of shoes, for which I now needed a size 15-AA. Thirteen-C was the closest thing Camp McClellan had. My AGCT (Army General Classification Test) score was one hundred twenty-two, which probably was a little above average for recruits at that time. I met a young man overseas later whose score was one hundred forty-four while I heard of one fellow in basic training with us who scored only sixty. The army was less selective then than now as it was desperate for enlisted recruits to replace overseas war veterans both in the Pacific and in Europe.

The third day at Camp McClellan, I departed Montgomery on a troop train loaded with inductees and headed for basic training at Lackland Air Force Base in San Antonio, Texas. We stopped for about two hours on the Huey Long Bridge in New Orleans where I was introduced to drinking wine. Someone had seen a liquor store back up the track some distance, and several fellows took orders with cash and walked back to purchase bottles of whiskey and wine for the troops before the train moved on. With my own wine in hand, I began to investigate the effects of blood alcohol levels upon my feelings and behavior. Within a short time, I was pretty well-plastered and having trouble making the trip to the rest room and back to my seat while grabbing every seat back along the way for support. During one of these trips, I unintentionally vomited on a tall blonde Scandinavian soldier named Houck from Wisconsin. He took a swing or two at me before deciding it wasn't worth the effort. We would later be friends and beer-drinking buddies, living in the same barracks at Clark Field in the Philippines. Our train stopped again at Camp Stoneman, Mississippi, for additional clothing and other GI issues before continuing toward San Antonio. Camp Stoneman also could not supply my needed shoe size. And so my feet were clothed with my own cloth athletic shoes for a few more days until the army finally found a size 15-C combat boot. Two pairs of socks took up some of the extra space and my slender young feet just had to adapt to the extra wide boot.

One of the purposes of basic training, besides physical conditioning and the learning of very basic combat skills, was to make you understand that your own wants and desires were of little or no consequence compared to those who held superior rank over you. The teaching of these principles was entrusted primarily to Drill Sergeant Hicks who supervised much of our learning and attitude modification as needed. An important saving principle in all of this is that the recruit is not alone and has a lot of company who are all in the same boat. Fortunately, for inductees, basic training lasted only three months. All 100-120 men in my flight (platoon) lived together, about fifty upstairs and a like number downstairs, in a two-story wood barracks with unpainted, raw wood floors. A row of double-decker steel bunks lined each wall with footlockers on the floor by each. A common toilet and bath served each floor at one end of the building. The mess hall was several blocks away to which we marched in formation with our flight three times daily. In fact, we marched in the same manner wherever we went except to the toilet or shower. We were awakened each morning at 5:00 AM to the tune of the bugle's reveille. We had fifteen minutes for toilet, dressing, and being in place in formation in front of the barracks for inspection before marching to breakfast. It was lights out every evening at 10:00 PM.

As I stood stiffly at attention in front of the barracks during a morning roll call and inspection in the first week of training, Sergeant Hicks moved slowly along each line of trainees. He stopped briefly before each man while fiercely looking to find something amiss or out of place. "Put that hat on straight, soldier! Eyes straight ahead, soldier! When did you last shave, soldier?" To the latter question asked of me as the sergeant plucked some fuzz from my chin, I meekly responded that I had never shaved in my life. "Fall out and shave, soldier! You're keeping your buddies from chow. We're waiting right here for you." I rushed upstairs to the second floor, found the unused razor in my footlocker, quickly scraped some barely visible peach fuzz off chin and cheeks without help of soap or water, and rushed back to my formation in five minutes or less. No one laughed or snickered. No one dared.

We were told that good behavior would earn us weekend passes to go into town if we liked. However, someone always screwed up or some other event always postponed the earning of this pass until the final two weeks of training. A common such event was KP duty for our flight on Saturday or Sunday. This duty always required the flight's presence at the mess hall from 3:00 AM until about 11:00 PM peeling potatoes, washing and stacking food trays, steam cleaning garbage cans, scrubbing floors, and a variety of other duties designed to insure complete exhaustion by the time one could stagger back to his barracks and bunk. I well remember one weekend on which we knew we were not scheduled for KP and thought that we surely had earned that pass. To our surprise, a short arm inspection was called that Saturday morning after chow (breakfast). We stood at attention with trousers down as the medical officer examined each penis, occasionally requesting the owner to "milk it down" for signs of a discharge or infection. Immediately thereafter, Sergeant Hicks ordered us to fall in wearing steel helmets and marched us across the base to an open field where a large pile of egg- to fist-size rocks stood across the street from the administration buildings. We were told to move all the rocks by carrying them in our helmets to a new location across the street. Alternating teams of recruits directed street traffic as the rocks were slowly moved. There was still hope that we might finish in time to get passes by noon. But such was not to be the case. When the job was finally completed, we were informed of a change in command. The rocks must be returned to their original location. "To every thing there is a season, and a time to every purpose under heaven . . . A time to cast away stones, and a time to gather stones together" (Ecclesiastes 3:1, 5).

On the shooting range, I had hoped to qualify for the sharpshooter badge but settled for marksman with both carbine and pistol. From regular calisthenics, strength exercises, obstacle course, long marches under backpacks, abundant healthy food, and usually regular hours in bed, I grew stronger and tougher but not significantly larger in any dimension. And so it went for twelve long weeks. Graduation was a proud event as our flight and others' passed in review before the reviewing stand from which the base commander and his staff returned our salute as the band played and goose bumps ran down my arms and legs and up my spine. We were now bona fide buck privates in the United States Army Air Force. Many of us believed we had signed up with the promise of a stateside technical school for which we were qualified before going overseas. We all were looking forward to different assignments in new and then-unknown places.

A week at home on furlough after basic training was just right. Just enough time to be seen in my uniform and to visit and hopefully go out with Ginger, a sharecropper's daughter a half mile up the road from our farm. Ginger was a high school sophomore whom I had noticed in recent years was no longer a little girl. We had eyed each other several times from a distance, nothing more. Her father was a hardworking farmer who made good crops year after year with a hardworking wife at his side and a large family, all of whom attended a country Baptist church on Sundays. Children in my family had always been discouraged from interacting with any of our country neighbors, so I knew that my parents were not happy about my visiting at Ginger's home.

Nevertheless, I stopped to visit upon seeing her on her front porch swing my first day at home. She was glad to see me, I think, and I, of course, was my usual nervous self. The next evening in my daddy's car, Ginger and I headed for town and a movie at the Princess Theatre. That was about as good as it got in those days, and I was more daring than ever, with my arm about her shoulder at the movie and holding her hand as much as she would let me. After drive-in refreshments and back home on her front porch swing, we talked about trivial things, not really communicating but trying anyhow. Her parents' watchful eyes were on the other side of the front screen door although they had nothing to worry about. Holding hands or a hug around the shoulders were the intimate limits of Ginger's comfort zone. A long, wet kiss was not in the books for this furlough. However, we did see each other and visit at least twice more that week. We would see each other again and exchange letters off and on for several years.

Overseas Duty. My orders were to report back to Lackland AFB for reassignment and transport to Clark Field on the island of Luzon in the Philippines. If a tech school opportunity was to be in my future, it would have to be overseas. This thought did not bother me at all. I looked forward with excitement to crossing the ocean on a troopship and seeing something on the other side of Mother Earth. Early in September 1946, two thousand new soldiers sailed through San Francisco Bay passed the island of Alcatraz and under the Golden Gate Bridge into the wide Pacific Ocean on board the SS Marine Cardinal. This was a "liberty" ship, originally designed for cargo and now modified to carry troops. Below deck, there were compartments 1-5 from top to bottom and A-D from front to aft. Thus, there were twenty compartments in which approximately one hundred men bunked in each. Bunks were arranged one above another three or four deep and in rows separated by isles too narrow for more than one man to pass at a time. Mine was a top bunk in compartment 5D. Air circulation was not the best, and quarters quickly became stuffy and overheated when all troops were present there.

For the first several days, we progressed steadily at about fifteen knots or more than seventeen miles per hour. After a week, our rate of speed decreased noticeably, and the monotony of daily routine began to set in. We later learned of trouble in the engine room. Our first brief stop would be in Honolulu where we passed the hulls of sunken U.S. Navy ships yet present in the harbor. We departed Honolulu still at reduced speed. Upon arrival at Okinawa, a typhoon was brewing that we had to ride out at sea for the next thirty-six to forty-eight hours. Troops were ordered to stay below deck during the storm, but compartment 5D was too hot and stuffy for all to comply with that order. I climbed the steel ladders to deck several times and hung on to a metal railing to keep from going overboard while getting some much-needed fresh air. It was a frightening sight to stand, as some of us did, on deck at the rear of the ship and see the front of our boat dive down into the trough of a wave towering seventy to eighty feet above it. When the ship's bow rose in the next moment to the crest of that wave, the incline of the deck was so steep that walking forward would have been impossible. As our craft was proceeding at an angle into the waves, it would literally twist in one direction forward and the opposite direction aft as if it might break in two pieces at any moment. When the storm abated sufficiently, we returned to discharge some troops at Okinawa before proceeding to Manila. Seasickness was common before and after the storm. I finally succumbed for a day or two after leaving Okinawa.

Bath showers on board ship were with cold salt water. Regular bath soap doesn't lather in such water. Of course, the military had a solution for that called "salt water soap." It looked and felt somewhat like the Octagon laundry soap used by my mother when scrubbing clothes on a washboard or boiling them in an iron wash pot. However, it didn't lather at all. After three weeks at sea, my scalp felt dirty and itched all the time. In desperation, I tried to shampoo my hair with this salt water soap. The result looked like hair covered with thick molasses that could not be washed out, and the sensation was worse than ever. I could hardly be still for scratching and would never have been able to sleep with that sticky mess on top of my head. A quick trip to the ship's barber for a buzz cut as close as the clippers could cut it gave me blessed relief.

We arrived at Manila harbor twenty-eight days after leaving San Francisco and traveling roughly nine thousand miles at an average speed of thirteen miles per hour. We were ready to feel earth under our feet, but to our dismay, it was mud six inches deep in the walkways and in the tents of a large tent city serving as a point of arrival and departure for fresh as well as tired troops, some arriving and others departing from Luzon. There was the customary waiting period of several days, as always in the army, during which we walked back and forth from our tents to mess and the outdoor toilets. Occasionally, a sergeant would appear with a list of names to call out and with orders to get on trucks to go perform some chore or other. One day, a few of us were driven several miles away to walk guard duty around a fenced enclosure in which Japanese prisoners stood around in small groups talking, smoking, and grinning broadly at their guards through the fence.

Clark Air Force Base, also known as Clark Field, is in Luzon, one of the larger Philippine Islands and several hours' drive north of Manila, the largest and capital city of the country. It certainly is not sufficiently north to escape the year-round tropical heat. Our barracks were one-story buildings with tin roofs, concrete floors, and walls of woven bamboo bark, interrupted at about waist height by a continuous window of screen wire with an outside metal overhang to block blowing rain. Inside, each soldier had a metal cot with mosquito net, which was an absolute necessity. An ammunition box standing on end beside the cot served as a clothes closet, beside which I managed to find a rough plywood table with two-by-four legs. Above this, I rigged a lamp to hang from a smaller empty box that served as a storage shelf. This was home for the next nine months until the Twenty-fourth Squadron of the Sixth Bomb Group of the 313th Bomb Wing of the Thirteenth

Air Force was closed down, and personnel transferred to other stations.

Meanwhile, there was plenty to do: regular daily assignments, periodic night guard duty on a long airstrip with B-29 bombers parked on each side of a runway for almost a mile, and one's regular turn at KP in the mess hall. Night guard duty on the flight line was even less desirable than KP. Regular application of mosquito repellent was necessary while the popping and cracking noises, made by metal aircraft cooling from the daytime tropical heat, helped keep you awake as you watched and waited for Hukbalahaps or Huks. The Huks were an armed peasant organization that had fought against the Japanese as U.S.-supported guerillas during the war. Now that the war was over, the Philippine government, dominated by wealthy landowners, wanted to disarm them. The Huks were reluctant to comply. General Douglas MacArthur cooperated with the Philippine government by having two of their top leaders imprisoned. Thus, by 1946, the U.S. military was no longer on good terms with the Huks, and the Huks were battling the Philippine army in several areas of Luzon province. All I knew of them then was that they were Huks, and some nights, we could see the flashes of light and faintly hear the explosions from big guns on the distant mountainsides.

Several weeks after my arrival at Clark Air Force Base, orders were given for troops from several squadrons to take up positions in the field on the periphery of the base and toward the mountains with carbines, helmets, and backpacks containing field equipment and supplies, including trenching shovels. My group was ordered to dig a trench with our pitiful little shovels in hard clay soil that a farmer's plow could not have penetrated. We made some progress but not enough to give us much protection from enemy fire. We spent a full day in the sun and were sent back to our barracks only after dark. As a private, I would never know if this was only some sort of training exercise or someone's response to a perceived threat that never materialized. However, I still remember the prevailing spirit of a group of young men, most of who had never been in battle; it was primarily of anticipation and excitement regarding the Huks and anger toward the stupid s.o.b. that would send us to dig a trench in that sunbaked clay with those little shovels. I'm sure that if bullets or shells had begun to fly, we would have learned something about genuine fear.

Upon arrival, many of us first served only as general-duty soldiers, but gradually, MOSs (Military Occupation Specialties) were assigned. Eventually, about a dozen of us were ordered to report for instruction at a radar mechanics school. There were only several hours of lecture on the basic science and technology dealing with the nuts and bolts of radio and radar equipment. The instructor was uncomfortable lecturing, read straight from the manual, and didn't like to answer questions. He seemed unable to comprehend that everyone did not already know what a tube and resistor did or how a radio circuit functioned. By the second day, he and most of the class were anxious to get on with altogether hands-on instruction. I knew how to hook a mule to a plow and, by now, how to turn on a radio, but I had no idea about what went on inside that little box. Therefore, the hands-on instruction was a waste of time for me. I soon let it be known to the sergeant in charge and was sent back to headquarters for reassignment.

My next assignment was that of mechanic's helper on the B-29 flight line. There I learned to dip various engine parts and other things in diesel fuel for cleaning purposes as well as how uncomfortable one could get on that runway when covered with grease and diesel fuel under the full glare of the tropical sun. My main qualification for this job was my ability to endure hard work and discomfort as well as anyone there. After six or eight weeks of this, I was transferred to work as an assistant clerk in the supply office under command of Lieutenant Lawford.

While at this post, orders came to prepare for closing down the Sixth Bomb Group. Among the many activities required was that of accounting for all issued equipment and packing some of it, including extra sheets and bedding for the entire group of several thousand men. Amazingly, it was found that a few thousand bedsheets were missing. How many might have been in use in native huts in nearby villages could only be a guess that there was no need to pursue. Under orders from Lieutenant Lawford, and with his personal assistance, we began to tear sheets in half and refold them to look like whole sheets to account for a couple of thousand missing sheets. I was stunned to find that an officer in our country's military would solve this kind of problem in this way. It was disappointing to me that a more honest solution could not have been found to the problem. After a while, I was promoted to the rank of private first class, not due to any particularly meritorious service but simply for having been in service for six months without screwing up in any major way. Shortly thereafter, a couple of my buddies from the radar school class were staff sergeants with three stripes and a rocker. I was envious but recognized that they had earned it and probably deserved it, and I didn't.

However, all was not work and disappointment. There were moments of excitement and entertainment as well. Five or six of the boys in my Twenty-fourth Squadron, including me, liked to get passes on the weekend to go together by bus to the nearby town of Angeles to drink beer in clubs operated by the natives for the entertainment of airmen at the nearby base. The native folks lived mostly in bamboo huts surrounded by rice paddies and fields grazed by water buffalo. There were pretty girls in many places, and Angeles was no exception. Even if some of us weren't dancers, these girls would stand to swing and sway or smooch with us on the floor in return for the purchase of another drink. I rarely did this due to shyness but also from concern for cleanliness of the girls about whom you had to wonder how they bathed and how much of their real body odor was camouflaged by perfume.

One Saturday evening, after too many drinks and missing the last bus to the base, four of us headed back to base walking along the dusty road. Along the way, there were flooded fields in which water buffalo grazed under a bright moon, quietly swatting flies or mosquitoes with their tails and splashing with their feet each time they moved. I recalled riding Maud in the pasture back home and decided that a buffalo should be much easier to ride. Ignoring the water, I waded out toward the nearest animal that paid me little attention. These animals were used for both work and milk and so were accustomed to being approached, though perhaps not often by drunken soldiers. The cow was startled when I jumped on her back, and she began to run, splashing as we went. I could hear my friends being entertained from the road, but my full concentration was needed to stay on the cow's back. And that was indeed rendered more difficult by the stiff needlelike hairs on her back that were penetrating my trousers and underwear, much to my discomfort. Suddenly, I had had enough and jumped off, falling headlong into the water. The next day was Sunday, and on this day, I did sleep late.

Ken Leech and I had become good friends on the boat from San Francisco. He was now stationed in another squadron nearby and flew regularly to get in his flying time as one of three waist gunners on the old B-29 bombers. These planes had all been through the war and now were not always in the best of shape. Ken had asked me several times to come fly with him and his crew on a weekend just for kicks, assuring me that his captain would have no problem with it. So one Sunday afternoon, I joined them, and we took off with me sitting in the midgunner's chair looking out the top blister as the only guest on board. One of the tires blew out on takeoff and became a matter of some concern.

We climbed to about five thousand feet or more and cruised down to Manila before turning and heading home. One of the crew noticed a fuel leak from engine number 3 requiring it to be shut down. Then the pilot became concerned about the performance of another engine and ordered us to don our chutes in case of emergency. Until that moment, no one was wearing a chute, and I had never put one on in my life. There were three of us in the waist compartment of the plane but only two full-size parachutes and one small chest chute. The sergeant present strapped on his standard chute and then strapped the little chute about the size of a rolled-up blanket onto my chest. The captain opened the bomb bay doors and ordered us to stand on a two-by-twelve-inch board in the bay area looking down at the rice paddies below just in case we should have to jump. The sergeant was anxious that I understand that I must not jump feet first but must tumble so that when the chute opened, my shoulder blades would be toward the ground to prevent the lines from striking me in the face. I must have turned as green as the rice paddies below. Thankfully, we never had to jump and landed safely in spite of a flat tire. I never volunteered to fly again just for fun.

Ken and I with two other fellows wanted to see Manila for a three-day weekend. In the tradition of old soldiers and sailors, I having passed my eighteenth birthday now, we agreed that we wanted to get "screwed, blewed, and tattooed." One Friday morning, with three-day passes in hand, we took a bus to Manila arriving on Rizal Avenue in the old midcity area by afternoon. I believed that Rizal Avenue once was the Philippine equivalent of New York's Fifth Avenue. There were still many blocks of building remains from wartime bombing raids. In some cases, the outer walls of a building of several stories height were still present. But on stepping through the front door and looking up, one could see the sky through space where floors had been, and everywhere, there was broken glass underfoot. I had no idea how much of that damage had been done by the Japanese and how much by American weaponry but presumed there had been door-to-door fighting as well as shelling and bombing. There were areas with less structural damage in which business was being conducted in food markets, dingy restaurants, various shops, and even a few hotels and nightspots here and there.

We found a hotel room with two double beds, a table with four chairs plus two worn upholstered ones, running water, towels, and wash pan. Toilets were down the hall. We dined that evening in the hotel restaurant, which was below ground level like a partial basement, not well-ventilated, smoky, crowded, and noisy. None of us were familiar with the menu items, so we looked at what others were eating and pointed to make our wishes known. I noticed several people ordering milk with their dinners and decided to follow suit. It was interesting to learn that the milk was served warm, not cold, and was caribou milk, not Jersey or Holstein. Nevertheless, I drank it. After dinner, we walked several blocks before wandering into a club where there was music and drinking and few Americans but numerous Filipinos. A native man with greasy wavy black hair, wearing a necklace and white suit, approached our table in a friendly manner to warn us that American soldiers had been waylaid recently after leaving this club. We had no idea who he was, why he might tell us something like that, or whether he might just be amusing himself. At any rate, we took turns going to the rest room and transferring our larger bills from our wallets into our shoes. We sat and drank beer and watched the "goucks" dance and party for a couple of hours before heading back toward our hotel. (The word "gouck" was a rude, slang, and commonly used term by soldiers in the Pacific region to designate dark-skinned persons native to that part of the world.) We were followed, but for only a block, by four goucks who we hoped were not carrying guns. Otherwise, we thought we might be able to handle them if we had to.

The next day, we found a tattoo artist on the street with his paraphernalia all around him and just looking and waiting for boys like us. He seemed old enough to have been in the business for quite a while and to be experienced in the trade. His obviously-used-but-not-so-clean instruments further testified to his long experience. He showed us a number of patterns and colors that could be used to trace pictures on our anatomies, and we began to think long and hard about what we really wanted to do. I had wanted a tattoo for some time, but upon seeing his needle still clogged with paint from the last job, I was having second thoughts. I conveyed to him my concern about his needle, and he began to clean it up while assuring me that all was safe. Ken wasn't overly concerned about the needle and wanted a large dragon on his upper arm from shoulder to elbow. The other guys made up their minds also, and I finally decided upon a small scroll with the word "Manila" printed on it, a blue bird above it, and a cluster of leaves and flowers beneath. It took a while, but we finally left with our new wounds greased with Vaseline and wrapped in toilet paper.

That evening, we encountered a pimp who wanted to provide us with "beautiful girls" at our hotel. We agreed to pay for four girls, 10 percent down and the rest on delivery. After a while, the pimp returned with an old man and three young girls in tow. To our surprise and dismay, they appeared to be children from eleven or twelve to maybe thirteen years of age. The whole scene was sickening. Ken and I refused to pay anymore. The other guys paid but, after half an hour, felt sorry for the girls and sent them away without ever engaging in any real sexual activity with them. And so our weekend goals were only partially fulfilled. We were indeed tattooed, and in an unexpected manner, we were "screwed." But that's as far as it went.

After nine months in the Philippines, I was transferred to the Twentieth Air Force and Guam in the Mariana Islands, where I would spend the next fourteen months. My friend, Ken, also went to Guam but to another squadron on the north end of the island from which B-29s were still flown regularly. I was sent to the Seventeenth Communications Squadron high on a bluff above Tumon Bay and Beach, which today is the site of hotels and tourist facilities. After three months of on-the-job training, the MOS of teletype operator was assigned to me with responsibilities for operating teletype or other kinds of telegraphic typewriters as well as telegraph printer switchboards for interconnections with other teletypewriter stations. I was cleared for handling classified information among and between military commands all over the world and, after three more months, was promoted to corporal. That was the last promotion I would get. If I had ever earned a third stripe, I might have stayed for twenty years or so.

Life on Guam for an American soldier at that time did not offer much variety. Off-duty activities most commonly included washing clothes, playing cards in the barracks or Ping-Pong at the USO club, swimming in Tumon Bay, and drinking beer at the service men's club between 6:00 and 10:00 PM. You could spot many of the old-timers, who had been in service for a decade or more and were making it a career, by the size of their waists as they sat on bar stools increasing their necessary belt lengths with each swallow. It seemed that one of the main duties of government in maintaining troops in the Pacific at that time was to keep them supplied with an abundance of beer. I did my part to consume my share, and fortunately for me at that age, it had no discernable effect on the size of my waist. There were some weeks when my comrades and I drank beer at the club every night of the week, returning to the barracks to sleep and start anew the next day.

On one of those evenings, when returning to the barracks by way of a worn trail through an open field, we approached a group of Latino soldiers talking around a small bonfire. These boys often stayed and played together and, on this night, were pretty high and looking for trouble. It was soon apparent that at least one of them intended to block our path so that to proceed further, we would either have to fight or detour greatly, which in itself might aggravate the situation. For some reason, he decided to focus his attention on me with his now-drawn switchblade knife and an increasing stream of profanity. My group, now walking about five abreast, stopped a couple of steps from him. His buddies and mine stepped back a little to form a sort of circle as he continued to brandish the knife and curse me in Spanish. Finally, I cursed him and told him to do his best at which he lunged with his knife. I instinctively responded with a lucky kick, dislodging the knife, which was picked up by one of my friends. We circled each other a time or two more, surrounded by a circle of onlookers, and then walked away and on to the barracks. I was by no means invariably a winner when physical combat was involved; however, I did learn that it is better to fight, win or lose, than to tuck tail and run. Because of this incident against the knife, I gained some respect from my peers that benefited my psyche considerably.

My last fight in the service was one in which I got my ears pinned back severely. It was a Saturday afternoon when I walked several miles to the barracks of another outfit to visit a friend from Alabama and basic training days. We were sitting on his cot and footlocker at one end of the barracks talking and listening to country music from a small record player. Meanwhile, a relatively older soldier and staff sergeant dozed on his cot at the other end of the barracks until he was disturbed, probably by a fly or some such thing. He barked for us to "turn off the goddamned music." We turned it down and continued to talk in low tones. He barked again, to which I replied that we had tried to please him but had no intention of turning the music completely off. He was up in a flash, rushing toward me and grabbing me by the collar at which I began to throw punches in self-defense. We fought out the door of the barracks and into the yard where he caught me with a punch that put out my lights for a few seconds. I woke up lying on the sidewalk with him standing above lecturing me about my foolishness for thinking I could take on a real man and combat veteran like him. I struggled to my feet and started swinging again before he knocked me down once more. I stayed down this time until he walked away.

After nine months on Guam, I was offered a week of R&R in Japan, which I happily accepted in early 1948. I can barely remember the flight from Guam to Tokyo. It took several hours and must have been aboard one of the old transport C-47s with a long board for seats on each side of the cargo compartment. The Emperor's Palace in Tokyo, with its beautifully manicured grounds and high iron fence and gates, was impressive as was the extreme order and cleanliness everywhere one looked. This was true not only of the cities but also of the countryside. It seemed that every square foot of ground was being used and carefully maintained. Nothing was missed, and nothing was wasted. Traffic seemed to move automatically as if controlled by a master switch. Neatness and organization was the order of the day in every place that one could see.

The train ride of several hours from Tokyo to Gamagori, a seaport city, took me passed snowcapped Mount Fuji in the late afternoon. The Air Force had made all arrangements. All that other service men and I vacationing there had to do was to be in the right place at the right time. I remember being conducted to a Pullman-type sleeping car in which the tiny little bunks were in size reminiscent of those in compartment 5D below deck on the SS Marine Cardinal. Also, when I stood straight, my head almost touched the ceiling of the car. The short conductor was so amused at the sight that he could not control himself and continued to laugh. He seemed overcome by both comedy and embarrassment. We arrived in early morning fog at a train depot in Gamagori from which we took a small bus to a nearby village in the mountains near the Nishiura Hot Spring. There were several small hotels as well as tourist shops, a few restaurants, and a public hot bathhouse that I was not inclined to try. My hotel and transportation expenses were already covered so that my only need for cash was for shopping, eating, or other entertainment outside the hotel.

On the second day at breakfast, I met and became friends with a young sailor about my age and an older army sergeant with close to twenty years in service. We shopped and explored the little village and nearby sights of interest and, on one occasion, dined away from the hotel in a restaurant's private dining room where we were waited upon individually by different waitresses. We visited a Shinto shrine where each of us was photographed in the priestly garb of the indigenous faith of the Japanese people. We drank sake, played cards, relaxed, and generally enjoyed nature's beauty and our free time without responsibilities.

While shopping alone one day, I met a little teenage sales clerk not much taller than my belt buckle. She was attractive mainly by virtue of her friendliness, although there are many very beautiful girls in Japan. I had purchased a little English-Japanese book of translations for tourists containing such phrases as "where is the bathroom?" that translated is "benjo doko wa deshka" or something very similar. She very much enjoyed helping me pronounce the Japanese correctly amid much laughter in the process. She insisted that I come to her house for dinner one evening and walked with me a couple of blocks from the shop to show me exactly where to come. I arrived with a little gift and with the hope it might be appropriate. From her response and that of her mama and papa, I presumed that it was. She reminded me to remove my shoes inside the door before walking on the straw carpets that were throughout the house.

My little friend's home was the only native Japanese home I was ever in. It was quite small and generally devoid of furniture by our standards. In the dining room, there was no table, and we sat on the floor with cushions if we liked. In the bedrooms, there were no beds, but pallets were rolled up and stacked in the corner. We ate rice and bite-size pieces of meat with chopsticks besides bread and tea. I brought my tourist book of translations for conversation, and everyone, including her mama and papa, seemed to enjoy the evening with considerable laughter before, during, and after dinner. Every time my bowl was empty, Mama would refill it in spite of my objections until finally I could literally hold no more. About a half hour later, Mama and Papa retired for the evening, and my little friend and I visited as usual for another half hour before we hugged and parted. The day I left the village to return to Guam, she was waiting at my bus, when I came to board, with a little present wrapped in brown paper. I cannot now remember either her name or the identity of the little gift so carefully wrapped. But I still remember her.

My only experience with a hot shower during two years overseas was at the hotel in Japan. Back on Guam, the weather was tropical, rainy, hot, and humid year-round. Unheated showers for the entire squadron were located in a single shed about fifty yards from our barracks at the end of a rocky dirt path. Although the water wasn't heated, the air usually was hot enough to make up for that. We stumbled back and forth from the showers with towels around our waists and wood clods on our feet. The climate was conducive generally to all kinds of fungal growth, including one similar to the athlete's foot fungus but more aggressive and troublesome. My feet probably contracted the "jungle rot" fungus from the always-moist concrete floor in the laundry and shower shed. I started treating it morning and night with regular GI foot powder, but to no avail. Within a few days, my toes were stuck together with scabs of blood and pus, and my feet were badly swollen. After a week's relief from duty with daily sick call at the dispensary, I began to heal. Thereafter, I was careful to dry thoroughly and use plenty of foot powder.

Stateside Duty. After twenty-four months overseas in the tropics of the Pacific, I returned to the USA in August 1948 and was reassigned to the Twenty-eighth Communications Squadron at Rapid City Air Force Base in Weaver, South Dakota. During a brief furlough at home in early September, I was shocked to learn that my favorite aunt Mae had died two months earlier and I was not told about it until then. Ginger was now a high school senior, and I went to see her right away. We went to a couple of drive-in movies and had a few milk shakes to boot. Morgan County was as "dry as a bone," and she would not have imbibed anyhow. Her parents were not rude but didn't have much to say and, I felt, were not thrilled to see me in all probability. Nevertheless, we visited quite a bit on her front porch swing. Also, my brothers were home for a day or two from college, where Robert was now captain of the Auburn wrestling team. Some minutes after exchanging polite greetings, he asked if I would like to wrestle with him in the yard. I agreed to a bout with him, feeling some uncertainty about the outcome. He got on his knees and gave me the advantageous superior position. Someone said "go," and in less than thirty seconds, I was on my back, pinned to Mother Earth. I tried to save face by telling him that was not the way I fought, but I knew for sure that the days of physically intimidating that little brother were a thing of the past.

After a week, I was back in South Dakota pounding the teletype keyboard and waiting, unknown to me at the time, for the arrival of the worst winter blizzard in U.S. history. My duty assignment there was more like a vacation than anything I had ever known. The communication center was small and, in the evening, required the services of only one operator. For the remaining eight months of duty, I manned the communication center every third night from 6:00 PM to 6:00 AM with lots of spare time on my hands. Since several of my friends had similar schedules, three of us decided to pool our efforts and started a window-washing service for hotels in Rapid City. We washed the insides and outsides of all windows monthly for a couple of hotels, the tallest of which was about ten floors high with a restaurant on the second and ballroom on the top floor. It took the three of us two to three days to do this job for the price of $150, which we divided equally. The restaurant and ballroom windows required hanging outside the building by a harness with feet braced on the wall or windowsill while scrubbing with a wet, soapy brush at the end of a long handle. I dreaded my turn hanging out with the harness and was always reminded, when I did, of the rice paddies several thousand feet below the open bomb bay of that B-29 in the Philippines. Fortunately, I suffered no accidents and had only one close call, which could have seriously injured someone else, when the scrub brush fell off the end of its handle and narrowly missed a pedestrian's head below.

The air force base had its basketball team with which I practiced a few times and dressed out for several games. Our player-coach was our point guard and a pretty good player. I was one of the taller ones, but served mainly to give others a rest. One evening, we played a very athletic-looking team of fast breaking Sioux Indians, called the Rosebud Legion, in an out-of-town game at a country gymnasium. There was a small crowd of spectators, mostly Indians. At half time, both teams used the same dressing room. When many of their players drew out bottles of "firewater" from their bags, I figured that we must now surely have the game "in the bag." However, such was not to be. We lost by at least twenty-five points with the fast breaking Legion running faster each quarter.

I had now passed my twentieth birthday eight or ten weeks ago, and two great mysteries of life, about which I still knew practically nothing, were girls and sex. I am sure that somewhere, there were soldiers equally ignorant of these subjects, but I felt that I was in something of a class to myself. Sgt. Randy Ramirez was a handsome nearly thirty-year-old veteran airman who was generally quiet, self-assured, knowledgeable, and well-liked by both men and women. He went sometimes with our group drinking beer at the club on base or to the bars in town. Waitresses would always notice Randy before anyone else. He just had a charming manner that seemed to draw women like flies to sugar. He had been married and divorced and now had a girlfriend in Deadwood, about an hour's drive to the northwest into the Black Hills. For some reason, he took a special liking to me that I appreciated but didn't understand.

Randy commonly drove to Deadwood once or twice a month on the weekend to spend the night with his girlfriend, Millie. One weekend, he asked me to go with him to see the Black Hills and to let him show me around historic Deadwood. He said there would be no problem with me staying at Millie's place overnight. The little town was a tourist attraction then and is today offering gambling, tours to numerous historic sites, and shows recreating life from the old Wild West. In 1948, it was a small place with the business district contained primarily in one street several blocks long with mostly two-story buildings on both sides of the street. Most businesses appeared to be saloons, restaurants, and places to gamble. A particular saloon called Aces and Eights advertised with a chair in the window and with red paint running down its back against which Wild Bill Hickok was said to have leaned one time too many. There was an old theatre long unused, except for tourists to view, with stage and bar at the top of the hill. The old dusty half-drawn curtains of red velvet fringed with gold gave it an air of authenticity and mystique. Standing there and looking, you could almost hear and see the music accompanying the singing and dancing showgirls striving to be heard over the boisterous talk and cursing of horny, thirsty miners from the nearby gold mines at Lead.

Most places of business were separated from each other by stairs that opened to the street and sidewalk below and to apartments on the second floor. In the late afternoon, after seeing the sights in Deadwood, we ascended one of these stairs and rang the bell before being admitted by an attractive, scantily clad young woman. She told Randy that Millie was occupied but would be out shortly and for us to come on to the kitchen for drinks or coffee or whatever. In the kitchen, I was introduced to a late fortyish or early fiftyish lady with auburn hair, pinned in a bun on the back of her head and wearing an apron. She was preparing dinner but managed to continue chatting politely all the while. It was apparent that she was in charge of this place and had been expecting us. We soon sat down to dinner at a large table in the kitchen, where I ate nervously while mostly watching and listening to what might be considered normal conversation in many American families. Occasionally, the bell would ring; a young woman would leave and, after a while, return.

During this extended meal, Millie suddenly appeared all fresh and pretty and overjoyed to see Randy as they embraced. After a few minutes, they excused themselves, and I didn't see either of them again until morning. In the course of the evening, a couple of the women casually invited me to go with them to their rooms if I liked, but I thanked them and said that I was just visiting. They could have made great sport of me if they had been so inclined, but they were sensitive to my timidity and uneasiness, and in time, I began to feel more comfortable. The madam drank beer with me and poured coffee for us both several times as we sat at her kitchen table and talked until after 2:00 AM. She explained the important services rendered to mankind for centuries at least by those in the oldest profession on earth. She believed herself to be a potent force in reducing rape and murder and in helping many lonely and frustrated men to function more normally. She claimed to have been a member of the chamber of commerce at one time and to be a long-time friend and supporter of the local sheriff to whom she and others in her business regularly paid for their protection. I think we were both exhausted when she finally invited me to sleep on the couch, which I did.

The next morning, I had eggs, bacon, biscuits, and coffee with "the family" and was invited again by one of the girls to go with her to her room. This time, I accepted the invitation. The price was twenty dollars. The time was about twenty minutes, including washing and inspection. Each bedroom was equipped with medical cabinet and sink with hot and cold water. Careful inspection and washing of all male parts preceded any sexual activity. It seemed more like a doctor's appointment than a romantic rendezvous. I was impotent with fear and with guilt for the sin that I felt I was about to commit. This polite and sensitive prostitute was concerned for me and suggested that I might want to see a doctor. I knew then that my problem was not physical and had nothing to do with my groin but that it was entirely between my ears. When we left Deadwood later that day to return to base, I was still very much a virgin and very disappointed with the whole affair. As far as I know, Randy never heard about my experience, and we continued to be friends.

On January 2, 1949, a savage winter storm hit Nebraska, Wyoming, South Dakota, Utah, Colorado, and Nevada. Known as the blizzard of 1949, this was the worst winter storm on record to ever strike the continental U.S. Fortunately for me, I was not on duty that Sunday evening when the temperature hovered in the thirties with only light snow forecast. There was a big poker game in progress in the barracks, and I was one of a half-dozen players. As the evening progressed, reports from outside told of darkness with decreasing visibility, increasing snowfall and wind velocity, and falling temperature. The poker game continued through the night and next day as the storm continued, and all travel to duty stations halted. On Tuesday morning, the storm was still intensifying with shrieking fifty- to sixty-five-mile-per-hour winds and swirling snow. Many doors and windows were blocked with snow and no longer useful. High winds erased radio reception, and phone lines were nonfunctional. The mess halls had not opened since Sunday evening, and we could not have gotten there if they had. There had been some sharing of candy bars and snack foods, but we all were getting pretty hungry. Wondering how long it would last, it seemed it would be forever. By midday Wednesday, the wind and snow were decreasing. While the doors at one end of our two-story barracks were blocked with drifted snow, we were able to open those at the other end. We broke down the doors to our mess hall, directly across the street, and passed cartons of canned food from its pantry to our barracks like firemen in a bucket brigade. By Thursday the cooks were again cooking, and snowplows were clearing streets of snowdrifts from ten to twenty-five feet deep. There had been as much as forty inches of snow in many places, and there were drifts up to the eaves of two-story barracks.

Mayors and governors across the region called for a state of emergency. President Truman finally answered January 29 by ordering the secretary of defense to use all money and resources required for Operation Snowbound. Hay from Nebraska, Colorado, and Kansas was trucked to the air base in Casper, Wyoming, loaded on to B-24s and C-47s, and dropped over ranches across the stricken states to feed stranded livestock. Local homemakers prepared sandwiches for the crews who worked night and day. In February, miles of road were cleared, and people liberated from their snowbound homes. However, it was early April before all roads were cleared, and spring brought great relief. The following month, I was honorably discharged two weeks early on May 16, 1949, by applying two unused weeks of furlough to my service time.

7

UNDERGRADUATE COLLEGE STUDIES (1949-52)

If I had been promoted another time or two, I would have been tempted to stay in the air force and make the military a career, perhaps for twenty years or more. However, my twenty-first birthday was only four months away. The clock was ticking, and I had not yet accomplished anything of significance. My brothers both were making top grades in premedicine at Auburn University and the University of Alabama. My one year of college with a grade point average of 1.8 had been a sad joke and a waste of time and money. Fourteen miles north of Decatur was Athens College, a small Methodist Church school, which—it seemed—might be a good place to start for one who had not yet learned how to study and still had no idea what he wanted to do, except to gain some degree of academic respectability. The GI Bill would pay for tuition and books plus one hundred and fifty dollars per month for four years of college if I could stand it. I had saved a little money with which I bought a used 1947 model blue four-door Chevrolet sedan. Daddy and Mother agreed for me to live at home while going to school, provided I obey their rules for midnight curfew and no drinking. I agreed.

We now had a telephone at home, so I could call Ginger without having to go to her house. I got in touch with her right away, and we began to go to movies almost weekly for a while. I think she enjoyed going out with me and she liked to talk, but mostly of people and things about which I knew and cared little. Also, I became a little annoyed and frustrated that she just didn't like to get into heavy petting. I had met and talked to a blonde high school senior girl named Lisa, who lived in Decatur across Oak Street from Mrs. Draper, who was now the paid caretaker for Grandmother Liston, since Grandfather Liston had died and Mother was teaching school. Lisa didn't have much to say but did love to smooch. And so we dated regularly for several months until our lips were about worn-out. I even gave her my King College letter jacket, which she proudly wore to school.

There were only half a dozen buildings on the college campus at Athens in 1949, although the college was founded in 1822. Student enrollment was less than five hundred, and four of those students commuted daily from Decatur in a blue 1947 Chevrolet to attend summer classes. It was a twenty-five-minute drive from Decatur, plus another twenty minutes for me to drive from the farm to pick up my paying passengers—Bill, Janet, and Hilda—at their respective homes before heading for Athens. Janet and Hilda were at least sophomores or higher and studying library science and prenursing, respectively. Bill Clark was a recently discharged navy veteran taking advantage of the GI Bill to take some courses in accounting and business. Since I had already had a year of college, I thought I should register as a major in some specific field of study whether I knew what I wanted to do or not. So I indicated business administration as my major and registered that summer for Business Administration 111, Political Science 221, Psychology 211, and Bible.

I was proud of myself for making three As and a B in these courses but felt that I had not learned a thing from the business and psychology courses that could ever possibly help me in any conceivable way. So then for the fall semester, I changed to become a science major and registered for Zoology 111, Math 162, Spanish 111, and Bible. Also, I had proudly worn my King College athletic letter several times and, in response to questions, told a few students that I had played a little basketball in the air force as well. It wasn't long before the assistant basketball coach, Coach Stone, invited me to come out for team practice. I was flattered, and the Golden Bears had respectable records in all the sports, including basketball. My undeserved and short-lived athletic recognition on campus would soon be recognized for what it was.

Cigarettes at about thirty-minute intervals and a little drinking every month or so were now a part of my life, and I did not give up either of these habits at this time. Also, I had experienced far less team-playing time and coaching in my entire life than any of the other members of the team. Coach Smith played me for a few minutes in each of a couple of early season home games and several minutes in a road game at Troy State before demoting me to the B team. Although embarrassed, I deserved no better than I got in this matter and, after a few more weeks, stopped going to practice altogether. An important result from this experience was that I turned with increasing energy to my studies and began to develop an interest in the biological sciences.

Dr. Reinich was a tall large-framed elderly spinster of German origin with big horn-rimmed glasses in which the lenses were unusually thick. Her almost masculine voice with heavy accent and the glare of her eyes, magnified by her glasses, could be a little intimidating if you didn't know her or if she didn't like you. All nine of the students in her class waited, sometimes sitting on the cool concrete floor, for Dr. Reinich, with keys rattling on a metal ring attached to her belt, to descend the stairs into the basement where she taught biology lecture and laboratory daily

8-10 AM. She was extremely knowledgeable about the classification, habits, and morphology of animals and plants; and her enthusiasm was contagious. She could get so excited, while demonstrating the anatomy of a crayfish and explaining its life cycle, that she might spray you with a little saliva if you sat or stood too close. And such enthusiasm was by no means limited to crayfish. For some reason, she took a liking to me. She gave me a B in zoology lecture and an A in laboratory. She would occasionally invite me to her two-room apartment on the top floor of the administration building for coffee and cookies when I had a free period. Upon my return for the spring semester, I was grieved to learn of her death during the Christmas holidays.

My longtime childhood friend, Neal Speake, was home for a week from Sewanee—University of the South, where he was on a football scholarship. We had not seen each other for at least three years, and since alcohol was not bought or sold in dry Morgan County, we decided to go honky-tonking to Huntsville on a Saturday night. After having beer at several bars, we found on the edge of town a busy barroom and dance floor with jukebox, filling station, and gas pumps all in one. The place was jumping with activity with several couples at some tables and three or four unescorted ladies at others. My dancing experience was limited to standing and swaying with a waitress on the floor of a barroom in the Philippines. And so while Neal danced with an unescorted lady from a nearby table, I stood at the bar drinking beer and surveying the scene.

At this point, four army paratroopers, in starched and ironed uniforms with jump boots freshly shined, came in muscling their way up to the bar. They were an arrogant-looking-and-talking bunch and seemed bent on finding trouble if they could. After ordering drinks, one of them turned to me and said in an obviously belligerent tone: "Hey, Mack, how come you ain't in uniform?" I paused briefly, looking into my beer, before looking him in the eye and saying: "Because I'm not a goddamned sorry bum like you and your buddies." And before he could respond, I hit him with a right hook, sending him through two saloon doors onto the floor of an unlighted poolroom. I don't think I saw him again after that, but Neal and I were suddenly in a slugfest with the remaining three. The management ordered everyone out, and a woman screamed: "Some son of a bitch had to start a fight." By then, the fight was moving toward the front door and out of the building onto the gravel around the gas pumps. I was delivering and receiving punches to face and body when another paratrooper fell backward to sit on the ground and lean against a gas pump without getting up. Neal had caught him with a solid blow.

Suddenly, flashing lights appeared from up the road, and a siren could be heard. The fight ended as everyone headed for their vehicles. I had parked on the shoulder of the road; so we were able to get away before the police arrived. We stopped several miles away to use a filling station rest room where I noticed, upon looking in the mirror, that my white shirt and khaki trousers were covered with blood and that my upper lip was cut completely through and hanging down on one side with teeth showing through the cut. The drive back to Decatur General Hospital took about thirty minutes and almost another hour to get the attention of a doctor, who stitched me up to go home at about 4:00 AM with a large facial bandage and numerous cuts and bruises. On arrival at the farm, Neal got in his car and headed home, and I went into the house to face the music.

Daddy, with his walking stick, had been sitting at the dining room window for a long time looking for car lights to come down the bumpy dirt road. Mother was in bed in the log bedroom crying as if her heart would break. I was guilty of breaking my promise and felt terrible to see them so upset. Daddy put a real verbal browbeating on me, threatening to throw me out of the house that very day. Gradually, my situation evolved to the point that I would be given one more chance to live within the rules. If I broke them again, I would be out of the house and entirely on my own. They loved me but were unwilling to live under the emotional stress that my behavior was causing them. Neal came out to see how I was doing the next day. And a few hours later, I overheard Daddy telling Mother that, according to Neal, "he acquitted himself quite well in the barroom brawl."

Dr. Reinich was replaced by Don Williams, who had just earned his master's degree in botany from the University of Tennessee at Knoxville. I first studied botany under Mr. Williams, who was a pleasant and friendly instructor trying hard to please everyone in his first job after finishing college. Biology was becoming a field of increasing interest to me, and its appeal was due to more than subject matter alone. The things that I could imagine biologists doing were not things that would necessarily require a great deal of interaction with people. Collecting, classifying, studying, experimenting with and writing about animals or plants seemed like a much happier way to live than any job that might require selling, politicking, public speaking, or the like. Therefore, I began to think about the opportunities that might be available to a biologist. It was immediately apparent that most towns probably had few openings for biologists. In fact, some citizens of the city of Decatur would like to have gotten rid of a budding biologist named Dan Speake, who caught and caged snakes in his home and backyard even after a city ordinance had been passed, on his account, to prevent such activities. His neighbors were much relieved when he finally left home to attend Auburn University, where he retired years later from the position of director of Wildlife Research as emeritus professor of zoology and wildlife.

When discussing my options with Mr. Williams one day, he told me about the various departments of biological sciences on the Knoxville campus. It was interesting to learn that the study of insects alone could justify so much effort that many state universities had departments of entomology. Realizing that pest control companies were businesses that employed numbers of people, and not thinking about the salesmanship and politicking that might be involved in such a business, I decided to attend the University of Tennessee for the fall quarter of 1950 on Mr. Williams's recommendation. But first, I had to finish the then-current spring semester and ensuing summer term at Athens to complete a few more credit hours, including physics, which I did. At the end of the summer, I was happy about the fact that I had managed to earn a 3.5 average for work done at Athens and to improve my overall college average, including the freshman year at King, to 3.0.

My parents both expressed to me their happiness and pride in my newfound academic success at Athens, and I am sure also felt considerable relief to see me launching out (of the house) again on my own. I believe that Mother would have preferred to see me major in almost anything other than a biological science and Daddy could not understand why anyone would go to college to study insects. It seemed to embarrass him a little to explain to friends that his son was off to the university to study "entomology," which meant the study of insects. Both would have been happier had I chosen law, medicine, or the ministry. And I believe that even then, Mother's concern was based upon fear that I might lose my religious faith if exposed too much to "Darwin's nonsense."

Realizing that paying for dormitory room and board at UT in Knoxville would no longer allow me the luxury of an auto, I sold my blue Chevy shortly after arriving at the home of my Lapsley cousins, where I was a welcomed guest for a brief period until classes started and I was settled in the dorm. In 1950, the boys' dormitory was conveniently located under the football stadium; and my roommate, Harold Lane, and I never missed a home game. In those days, Doug Atkins and Hank Lauricella were royal knights of the gridiron on the hill in Knoxville, and the UT football team members were declared national champs my senior year. General Bob Neyland was coach, and team discipline was of a military style. If you fumbled the ball in a game, you would be seen carrying it with your books everywhere you went for the next week. The UT Volunteers were the only team in the Southeastern Conference still running every offensive play from the single wing.

However, dorm life was not altogether ideal. Our room was small with two small closets, two wood chairs, two wood desks, and a double-decker bed. At the end of the hall was a public bath in which a row of toilets lined one wall and urinals the other, with no private stalls. Similarly, a dozen or more showers lined the wall of a common stall at the end of the room. Whether sitting at your desk studying or sitting on a commode grunting, you could hear Scotty Rankin blowing his trombone from the other end of the hall in make-believe accompaniment to "Satchmo" Louis Armstrong. And if you sat on a toilet seat often enough, there was a good chance that you would eventually take crab lice back to your room. It was my misfortune to do just that and to suffer with the problem for several days before discovering the source of my discomfort. I had lost sleep for two or three nights in a row until in desperation, I turned on the lights in the wee hours one morning before sunrise to examine my pubic area with a magnifying glass. To my horror, I could see numerous tiny, gray, wingless, six-legged, broadly oval, somewhat crab-shaped insects crawling over raw skin among hairs to which tiny things, later found to be eggs (nits), were attached. I shaved some of the area immediately, placing the hair in an empty bottle, and was at the drug store when the pharmacy opened that morning. Laundering bed linen and proper application of an insecticidal salve solved the problem in less than twenty-four hours. Since *Phthirus pubis* is not known to transmit any disease, I was safe from that standpoint. However, after these experiences, Harold and I decided to rent a room off campus for the next term.

By the end of the fall quarter, I had learned that good grades would not be as easily come by here as they were in Athens. I had earned Cs in English literature, Spanish, and genetics and Bs in chemistry and entomology. I had carried more hours than usual but was also greatly distracted by all the constant extracurricular activities that then had characterized dormitory life. Scotty, the trombonist, was a party animal as well, and several boys, including me, had succumbed to his idea of celebrating a couple of weekends before quarterfinal exams. Although I couldn't afford the time or money, I went along with the group to a Saturday nightclub celebration in nearby Oak Ridge. The most entertaining thing about the entire weekend to me was watching Scotty jitterbug with half a dozen different women, spinning them around and turning flips like I had never seen done in person before. I faked my way through a couple of slow dances and sat and drank beer with the boys most of the evening.

Mrs. Borne and her husband lived on White Avenue only a five- to ten-minute walk from Ayres Hall, a multistoried classroom building in which some of our classes were held. Their children were grown and gone, and they had recently decided to rent an unused upstairs bedroom to college students. They seemed pleased with us and we with them. And so we agreed to pay one hundred dollars per month, fifty dollars each, with toilet and bath down the hall. We had two single beds, a desk with chair plus an easy chair and hot plate. Canned pork and beans were often heated on that hot plate for lunch or dinner. Harold was a navy veteran trying to survive on the same monthly GI Bill allowance as I. We walked everywhere we went, and I sometimes got soaked as I preferred running for cover to carrying an umbrella. This was home for me for the next eighteen months and until graduation.

For the winter quarter, my grades were three Bs and an A. And then in the spring, something happened that I have never understood but that was a serious blow to my morale. My grade report showed Bs earned in chemistry, literature, history, and Spanish, and what seemed to me to be an impossible D in insect taxonomy. In mentally reviewing my exam, I felt that I had done well. But I wondered if Dr. A. C. Cole had suspected me of cheating. During the exam, we were required to walk back and forth from our desks to a front desk in order to obtain specimens that we identified back at our seats under the microscope if necessary. My seat was in the back of the room. Therefore, I had to pass several other busy students in coming and going to the front desk. Could Dr. Cole have thought that I was spying on the papers of others? I wondered, and the very thought made me almost physically sick. I went immediately for a conference to his office, where I was allowed to cool my heels for a while before seeing him. He was head of the entomology department with a faculty of three. When I was finally allowed to confront him across his desk, he would only smile and say that I had made what I had earned. He would not tell me which questions I had missed or how much the final exam had counted toward the final grade. That was none of my business. It was all a matter of his judgment. This was my first course with Dr. Cole.

I still well remember my emotions while walking down the hill from that conference, eyes filled with tears. I swore to myself, somewhat as I had done a few years earlier at Camp McClellan that, regardless of what might be required, Dr. Cole's opinion of me would have to change. I had respected him until now but never liked him again. To me, he was an enemy who had to be defeated at all costs and courted only when absolutely necessary. When the course was next offered the following year, I earned an A in it along with As in three other courses that he taught. It was my policy from that time on, in every class of his, to sit in the middle of the front row, maintaining eye contact with him as constantly as possible, unless taking notes. On every exam, I made a show of waiting until he entered the room to place all books and notes on a windowsill, roll my sleeves above the elbow, and wait with clean paper, sharp pencils, and eyes on him for whatever he had to offer. He not always but often did ask essay questions requiring both specific knowledge and organization of thoughts to be written. I suspected on one of his exams in insect morphology that he might ask about the detailed structure of the ovipositor of an ichneumon wasp. The structure was rather difficult to envision even with the illustrations provided in Snodgrass's text. So, to be safe, I had memorized word for word an entire paragraph on the subject. Lo and behold, he asked the question! And I gave it to him, word for word, right out of Snodgrass. He never complimented me verbally but did give me an A+ on the test. It is difficult to describe the emotional satisfaction that I derived from that test.

There was an older fellow, whose name I don't remember, taking some of the entomology courses with me to gain background knowledge on insects in order to improve his standing with a pest control firm for which he worked. As he described his work on the job, it didn't sound that challenging or interesting to me. A young man in one of the senior courses had been awarded an assistantship for graduate study in entomology at Ohio State University. He seemed to have a cozy relationship with Dr. Cole. I liked Dr. Fred Lawson, another entomology professor, and started talking to him about possible graduate study in entomology. I had done well in his economic entomology course, and he had encouraged me a great deal. He told me about a good, nearby department of entomology at North Carolina State University.

In the winter quarter of my senior year, I applied to North Carolina State and Dr. Clyde F. Smith, head of the entomology department there, for an assistantship in entomology. Their response was a positive offer of an assistantship but on condition that I make all As in all courses in my last quarter at UT to help offset that ugly freshman year at King College and my first quarter at UT in which I had made only Bs and Cs. I met the challenge with all As in four science courses and graduated cum laude on June 9, 1952, with a BA degree in entomology and zoology. I had been going without sleep for weeks in order to study with the aid of *no-doze* pills followed by *sleeping* pills, and my weight was at least fifteen pounds below my usual at one hundred fifty-five pounds.

I have been reminded, then and many times since, of Daddy's words at the homework table on the farm many years earlier to the effect that he had often encountered opponents smarter than himself but never one that he could not beat through hard work. I am thankful to him for that guiding principle that has carried me successfully through many a challenge. He and Mother drove to Knoxville and attended my graduation exercises. Mrs. Borne and her husband insisted on having a celebration dinner for Harold and me and my parents in her dining room. She was a sweet lady.

For the past two years, I had done no dating at all in Knoxville. I had hitchhiked home for Christmas a couple of times, during which I visited casually with Ginger. We had exchanged letters several times at considerable intervals, nothing romantic, just newsy, friendly letters. Such correspondence helped ease my sense of isolation. There had been a girl in some of my biology classes at Knoxville with whom there was mutual appreciation and enjoyment while working together as lab partners in a microtechniques course or occasionally eating together at the cafeteria. However, I had neither time nor money for further pursuing such interests.

8

Graduate Studies (1952-57)

My parents and I drove the following day, June 10, 1952, from Knoxville to Raleigh, North Carolina, and the North Carolina State University campus. My first residence was in Watauga Hall, an old boys' dormitory located near the university library.

North Carolina State University. The following day, I met Dr. Smith in the basement of the old zoology building where entomology offices were then located adjacent to a greenhouse also used by entomology faculty and students. He told me that there was a real need for research on peanut insects and that we needed to know a lot more about one insect in particular, the southern corn rootworm (SCR), which also attacks peanuts. He said he would really like to see someone develop a technique for mass rearing this insect in the laboratory in order to support a variety of research projects. He suggested that I might want to consider taking on this SCR as the subject of my thesis research. That day, I was officially registered to pursue three quarter hours of credit in research during the summer term toward my master's degree and to be a graduate student assistant for a professor, Dr. Jim Dogger, who would soon arrive to assume responsibility for studies on the biology and control of peanut insects.

My assistantship paid one hundred twenty-five dollars per month, which, when added to my GI Bill allowance, gave me a total of two hundred seventy-five dollars monthly. And I still had three years of eligibility remaining on the GI Bill. Things were definitely looking up, but I knew there would be challenges ahead, the nature of which I knew practically nothing. I still had no definite long-term goals in mind other than the need to determine what my immediate research objectives would be and how I would go about achieving them. The idea that I was now being paid fifteen hundred dollars annually by the state of North Carolina to assist with important research and also to conduct an individual project of my own, though not yet clearly identified and which must somehow enlarge mankind's store of knowledge, were momentous responsibilities now weighing upon my shoulders, or so I felt.

Almost a dozen graduate students shared a basement room in the old zoology building, each with his or her own desk, chair, and improvised bookcase. Don Weismann, one of these students who later took a job at the U.S. National Museum as a taxonomic specialist on the leaf beetle family Chrysomelidae, showed me specimens of twelve-spotted cucumber beetles (*Diabrotica undecimpunctata howardi*) from his collection and suggested how I might catch a few and cage them in order to collect eggs that would hatch into little SCRs. I soon had a number of live beetles feeding on lettuce in a lamp chimney cage, closed at the top with cheesecloth and a rubber band, resting on moist blotting paper in a petri dish. After a few days, I was collecting tiny yellow beetle eggs from the blotting paper, with the moistened tip of a camel's hair brush, and transferring them to salve boxes of moist soil with germinating corn seeds. And so I began to spend time daily feeding the beetles, transferring eggs, checking larval development, and recording the time periods spent in each developmental stage. Daily records of maximum and minimum room temperatures were also recorded. It was suggested to me that I should review the life history of the insect from the literature as well as by rearing it myself. So, in addition to caring for my pet SCRs, I began to spend as much time as possible in the library searching for everything that had ever been published on *Diabrotica undecimpunctata howardi*.

My roommate was a junior in textile engineering, a handsome and athletic-looking fellow from New York who drove a new red and white Pontiac convertible. For Jeff, college was just a stepping stone to a good-paying job. He was not particularly fascinated by any of his courses but did what he had to do to make grades sufficient to satisfy his father. His friends were philosophically similar in nature and seemed to enjoy partying more than any other aspect of college life. He took me for a spin in his new car and got it up to one hundred twelve miles per hour before releasing the accelerator on a farm-to-market road with relatively little traffic. That is still the fastest I've ever traveled in an automobile and faster than I ever want to again. He and his friends, both male and female, were attractive, highly animated, confident, and sociable young people whom I admired in a way but was not really comfortable with. I double-dated with Jeff a couple of times and went to several of their parties before becoming too busy with other things to continue those activities. The daily experience of walking everywhere I went while my roommate drove about in his new car caused me to think about the possibility of again getting myself a car. And so it wasn't long before I located and bought a low-mileage 1947 model black four-door Kaiser sedan for an amount that I only remember as a bargain price. It looked something like a scarab beetle in form but drove well and gave particularly good gas mileage on the highway in overdrive.

The completion of a new life science building, Gardner Hall, was anticipated shortly, and the entomology department was scheduled to move into all three floors of one wing of the building.

The actual move took place early in the fall quarter of 1952 just as Dr. Dogger arrived to assume responsibility for peanut insect research. Suddenly, I had a small office of my own on the third story of the new building and next door to Dr. Dogger. I liked him immediately. I had just turned twenty-four, and he was perhaps a dozen years my senior. He was soft-spoken with a sort of shy sense of humor and a rigorous work ethic. Students on half-time assistantships could take only seven or eight hours of credit course work per quarter to allow time for work on their assistantships. Therefore, for one to several days each week, Dr. Dogger and I would be on the road to the research station in Rocky Mount, the Peanut Experiment Farm in Lewiston or elsewhere to work in field experiments or to survey fields for insect problems. These trips might require half days, whole days, or even weekends on occasion. We talked a great deal about everything from insects to general science to religion, including philosophy and sometimes even girls. He liked to tell how he has to remind his wife that she shouldn't get upset unless or until he stopped noticing pretty girls.

Besides Dr. Dogger, other members of my graduate committee were Dr. B. B. Fulton, Dr. T. B. Mitchell, and Dr. W. A. Reid. Dr. Mitchell was author of a book on bees of North America, was good at photography, especially of insects, and was chairman of my graduate committee. Dr. Reid taught plant pathology, one of my minor fields of study. Other classes from which I particularly benefited were insect physiology and toxicology courses with Dr. Robert Gast, immature insects with Dr. Dogger, and applied entomology with Dr. Walter Kulash. They and others contributed significantly to my education at that time.

Dr. Fulton was a grasshopper and cricket specialist, widely recognized for his knowledge of these insects and for his studies of insect behavior. He was also a musician, played the piano, and had published a key to the tree crickets of North Carolina that was designed to separate the different species on the basis of their songs. Eight or ten graduate students, of whom I was one, were privileged to take his graduate course in insect behavior the last time he taught it. On one of several field trips, we encountered a sphecid wasp in the act of provisioning her nest in an area of bare ground under some trees. He had us sit in a wide circle around the hole she had dug and told us to take detailed notes on everything that she did. She was in the process of dragging a small caterpillar down into the hole that she later sealed before flying away. We sat there for the better part of an hour watching and taking notes. At our next class meeting, each student was asked to read his detailed account of what the wasp had done. No two stories were identical. One said that she kicked dirt into the hole with her hind legs, another that it was with her front legs. One said that she packed the dirt with her head and another that it was with a pebble held in her jaws. It was a great lesson on the importance of careful and recorded observations.

Dr. Z. P. Metcalf was a world-recognized authority on the homopterous insects, which includes the aphids, leafhoppers, scales, etc. He was fascinated by the evolution of these insects, how their geographic distributions correlated with generally recognized floral and faunal regions of the earth, and the many implications of these studies. A staff of three ladies worked in his office to support his continuous correspondence, writing for publication, and maintenance of his large insect collection. For an entire quarter, eight of us graduate students attended his class on zoogeography each Tuesday, Thursday, and Saturday morning. Besides a hefty textbook of five to six hundred pages, which we read from cover to cover, we were given outside reading assignments totaling hundreds of pages. Without absorbing a significant amount of information from all the assigned reading, no student had a prayer of passing with an A or B that was the minimum acceptable performance for graduate students. Dr.

Metcalf met the first class of the quarter for the usual preliminaries plus several reading assignments. At the next meeting, he addressed the class by making additional assignments and asking if anyone had questions. As there were none, he dismissed the class with a reminder that there would be a written exam every Saturday morning. Thereafter, there were always student questions to fuel class discussion. He never lectured from notes, and what he had to say was usually guided by student questions. Needless to say, after that, we always had questions enough to fill the class period, and interestingly, he always forced us to help answer them by recalling things that we had or should have read. By modern teaching standards, he was a slave driver. I did make an A in his course due in part, no doubt, to my ability to write fairly well. However, I think I disappointed him later as he saw my interests drifting more and more toward applied or economic entomology and away from the more basic and philosophic areas of science. He interrupted me, as I read the *Journal of Economic Entomology* while walking down the hall one day, by asking "why are you wasting time with something like that, which will no longer be pertinent after a decade or two?" I answered to the effect that someone had to be concerned about such things, but he obviously was not impressed. Privately, I didn't see how the state or national economy could afford to support too many like Dr. Metcalf. His work was interesting but did not produce any goods or services that society depended upon.

My thesis research problem was well-defined by fall of 1952, and I was completely immersed in those activities plus other assistantship-related duties and formal course work. In my thesis research, I was seeking to answer as many questions as possible about the biology of the SCR as it related to peanut culture in North Carolina. How many generations occurred each season in peanut fields? When does infestation first occur? What factors contribute to the development of the insect and its appearance in large or small numbers? At what period during the growing season is most damage done? How important is this damage? To answer these questions, SCRs were reared in the laboratory and collected regularly from peanut pods and soil samples taken from the field. Peanut plants were dug up periodically to determine percentages of pod injury by SCR during the crop season. Beetles collected year-round from light traps were dissected to determine seasonal periods of oviposition. Known numbers of beetles were caged outdoors to determine winter survival. Drawings and photographs of insect stages, parasites, and crop damage were prepared for my thesis. Various kinds of data were regularly collected that would later be summarized for presentation. It was the kind of problem with which you could always profitably spend more time, and for which you couldn't imagine having too much data, and always wished you had a little more.

I drove home to spend a few days at Christmas with my parents. My brothers must have been there also for at least a day. I must have seen Ginger during that period. My clearest memory of the trip relates to driving my Kaiser through the Smoky Mountains on the way to Alabama. In those days, I could do fairly well without normal amounts of sleep, or so I thought. I left Raleigh after dark to drive over five hundred miles to Decatur, Alabama. This was before there were any interstate highways. It was cold outside, and I was getting very sleepy when I decided to lower my windows and turn up the volume on the radio. Some time later, I was awakened as the car bounced through a patch of oats dangerously close to the edge of a cliff hundreds of feet high. I stopped immediately, closed the windows, locked the doors, and slept for a couple of hours before completing the trip. The experience appeared to be pretty good evidence that my guardian angels were not napping.

Love and Marriage. It was some time in July 1952 when Ms. Chloe Hodge, the departmental secretary, invited all the graduate students in entomology and zoology to an ice cream party at her house on Poole Road several miles from the campus. She was a person whose company everyone enjoyed, and the students looked forward to this occasion that was not a first-time event for some of them. It was a Saturday afternoon when I drove to the party and parked behind several other cars in the driveway. In the backyard, a freezer of homemade ice cream had been opened and was being devoured. On the back porch, I met Chloe's brother-in-law, Oscar, as he wrestled to slowly turn the crank on another freezer. A third freezer had already been cranked as much as possible and was waiting under a moist potato sack for the hungry crowd. In the kitchen, I was given a bowl and spoon by Flonnie, Chloe's sister, before hurrying back outside to fill my bowl. I was just finishing my first bowl of ice cream when a tall suntanned girl with beautiful long legs parked her bike by the porch and went into the house. She was Chloe's niece, and her name was Janice. I suddenly felt an urge to use the bathroom and went into the kitchen to inquire of its whereabouts. This tan, hazel-eyed beauty looked at me with apparent disgust and said "it's right out there in the yard where you were." Weeks later, she told me that she had thought I was trying to embarrass her because they didn't have an indoor toilet.

Chloe Hodge was one of the essential parts of the entomology departmental machinery that enabled the wheels to continue to turn. She was the head secretary for the department and the department head's private secretary. Matters related to project funds and orders for supplies, equipment, and services all crossed her desk. She was always available for counsel or help with all kinds of departmental or even individual and personal problems. She typed most of the theses and dissertations required for graduation of MS and PhD candidates from the entomology department, plus many from other departments as well. She usually had a big smile and encouraging remark for everyone;

and most students and faculty enjoyed her company, wit, and sense of humor. I was no exception to this rule.

One October evening, I entered Gardner Hall after dinner as usual to work several hours in my office. Lights were on in Chloe's office where she apparently had stopped by for some reason. I paused for a drink at the water fountain and was headed for the elevator when her office door opened and her niece, whom I had met at the ice cream party in July, emerged for a drink as well. We passed each other with smiles and "hellos"; and I proceeded upstairs to measure SCR head capsules before turning my attention to books, assignments, etc. However, I began to think of other advantages and perks that might be associated with Chloe's friendship; and concentrating on work was not easy that evening as I continued to picture in my mind that beautiful white smile with full lips, hazel eyes, and gorgeous legs. I finally gave up before midnight to drive back to the dorm for a shower and bed. After all, tomorrow would be another full day as all of them were.

On a Saturday evening in early February 1953, Chloe joined Mrs. Thompson, a secretary in the zoology department, in treating the graduate students of both departments to a steak dinner at Mrs. T's home. Mrs. T's daughter, Sally, was there for auxiliary hostess functions as was Janice, who came with her aunt Chloe. The dinner was good as was the company and the music afterward. Several couples danced a little to a slow record, and I even ventured to invite Janice to dance. We barely moved our feet as we rocked back and forth to the slow rhythm of Tommy Dorsey or Les Brown and his band of renown or maybe it was someone else. Who knows? I was just thrilled to be standing so close to and touching that beautiful girl. So thrilled in fact that I called her the following Monday evening to ask her to double-date with me and Charlie Wright, who wished to take our dates dancing the next Saturday evening, February 14, Valentine's Day. She accepted as did Charlie's date, Velma Creech, and the four of us danced and visited at Club 15 until almost midnight.

Janice worked regular hours as a clerk in the state Department of Motor Vehicles. She and her girlfriends always seemed to be going to the movies or to a dance at the YWCA or something that they enjoyed and looked forward to. She had recently broken up with an old boyfriend. My work never seemed to end partly because that was almost the only thing I had ever done. One Saturday night, she invited me to go with her and her girlfriends—Dorothy and Betty Lou—to a YWCA dance, which I did. I started calling her to go to a movie or sometimes even to let me just bring her to my office to visit while I finished with the SCRs before going for a milk shake. She was always agreeable and ready to go. I was even able to collect insects from the lighted windows at the drive-in without her objection, although she did sometimes slip down low in her seat as if to avoid being seen. On a few occasions, she even accompanied me to the Rocky Mount Experiment Station to record data on a clipboard as I dug for and counted rootworms in a peanut field.

Dorothy and Betty Lou were members of the Presbyterian Church where Janice was known and often sang with the choir, although she and her family were active members of the Church of Jesus Christ of Latter-day Saints (LDS or Mormon). Worship services for the two congregations were not always at the same time. Dr. Smith, who was also LDS, called me in for a talk when he heard that I was dating Janice. The gist of his advice to me was that while there are many good things about all varieties of Christian faith, I should be aware that the LDS faith is more different from that of other Christian churches on the whole than they all are from each other. And that such differences, while not seeming important to me now, could become real problems down the road, not only between my present family and me but also between my possible future wife and me, unless we took the time to be absolutely certain about what we were doing. I liked Dr. Smith and appreciated his fatherly concern.

I started attending Sunday school and sacrament meeting occasionally with the LDS Raleigh branch in a rented room above the concession stands and dressing rooms at the Pullen Park swimming pool while their first church building was under construction and nearing completion. Oscar and I and the Barnes family often finished picking up empty drink bottles and sweeping the floors just in time for the first meeting to start. I was often invited to dinner with the Rogers family afterward and began to feel somewhat at home there. Aunt Chloe had always lived with the family and was the authority on matters of church doctrine, which she and I often discussed. I think it was about this time that I began to read, in addition to entomological publications, some LDS literature, including LeGrand Richards's book, "A Marvelous Work and a Wonder."

Janice and I were together several times each week now and talked daily on the phone. On the evening of April 2, 1953, while parked on the bank of a scenic lake near Raleigh, I proposed marriage, and she accepted. I believe we were both supremely happy at that moment and, with little interruption, for several months thereafter. Neither of us realized how thin the ice on which we skated was. When Mother learned in April that I was dating, her first question was to learn to which church this girl belonged. Words can't describe how distraught she was upon learning that the girl was Mormon, a member of a "sect" established by a false prophet, all of whom were in danger of eternal damnation for their heretical beliefs about the nature of God and Jesus Christ. She started reading everything she could about Mormon doctrine, and the more she read, the more upset she became. I received long letters, weekly until our marriage in September 1953, attempting to change my mind and to prove to me from scripture the serious doctrinal errors of that strange new religion. We drove to Decatur for a weekend for Janice to meet my parents during which Daddy was polite and tried to make Janice feel welcome while Mother had difficulty concealing her real feelings. Robert and Needham also came to visit and meet my fiancée.

Several weeks later, I returned for a brief visit during which Mother contrived to arrange a meeting between her son and the daughter of a friend of hers, but all her efforts failed. On June 25, 1953, I signed an agreement with a Raleigh jeweler to make six monthly payments on a small diamond ring. That evening, I put the ring on Janice's finger as we sat in the car by the state capital building before walking through the capital grounds to Fayetteville Street and the Ambassador Theatre. It never occurred to me then that a girl's birthday might not be the most appropriate time to give her an engagement ring. Long before this, I may have been programmed to "kill two birds with one stone" whenever possible.

As I was not an LDS member, temple marriage was not yet a possibility. Janice had always dreamed of a church wedding and had saved her money to buy her wedding dress. In spite of my mother's concerns and feelings, we wanted our families to be involved and were unsure how well Mother's constitution might tolerate attending her son's wedding in a Mormon Church building. And so, following a rehearsal party, sponsored the previous evening by one of Janice's aunts and uncles at their home, we were married September 11, 1953, at friend Dorothy's First Vanguard Presbyterian Church of Raleigh with her minister, Rev. Buffalo, officiating. My father was best man. My brother Needham and graduate student friend Rudolph Howell were ushers. Dorothy Simpson was maid of honor. Aunt Chloe and Sterling Weed, an LDS elder and graduate student in agronomy, sang prior to and during the wedding. The customary kissing of the bride and groom, which commonly ends a ceremony, was dispensed with as the groom was not comfortable with such intimate and public shows of affection. A reception, sponsored by the Raleigh branch of the LDS Church, followed across town at the new LDS Chapel, where members of Dr. Smith's family helped serve punch and refreshments. After handshakes in a reception line, I was ready to go anywhere just to get out of there.

I had no idea how offensive this behavior was to my wife or anyone else. I just wanted to go! My unwillingness to have the traditional kiss at the conclusion of the wedding ceremony and most of my behavior at the reception following were perhaps more revealing about the nature of my psychological hang-ups than any thing I had done in a long time. I could function moderately well in many one-on-one situations and sometimes in small groups. But when in larger groups and in full glare of the spotlight, I could become a social basket case. The wedding not only involved a large group with me being part of the center of attention but also kissing that is related in a way to sex, a subject with which I was not greatly at ease—certainly not in a crowd. In this situation, I could have mistaken a little innocent kidding for an insult and caused a real ugly scene. On this particular evening, I would have much preferred a gunfight at high noon to conversation and handshakes. Almost as frightening at this point in my life was looking ahead to a day when I would have to speak at a graduate seminar.

And so we left the reception with me driving our black Kaiser and Janice still in her wedding gown in the backseat. We drove back to her home on Poole Road with her friends—Bill and Katy McNally—following to take pictures of her in "going-away clothes" and the two of us going away. Janice had no idea what kind of surprise awaited her on her honeymoon trip. Of course, she knew we were leaving town in the Kaiser rather than flying, and since it was already evening, she knew perhaps that we would not get far before having to stop for rest. I would learn another day how disappointed she was when I pulled up and parked in front of the Knotty Pine Motel on the outskirts of Raleigh. The rest of that evening may best be summarized by saying that I had few good ideas and no useful training or advice on how a man should behave on his wedding night.

The next morning, we were cheered to learn that my application for married student housing on the North Carolina State campus was not made in time to secure an apartment right away. Therefore, we spent the day looking and finally found a small furnished upstairs apartment on Blue Ridge Road near the fairgrounds. About 5:00 PM, we drove back to her parents' home to load her belongings into our car and her mother's before returning to our new apartment for our first Saturday night at home. On the way, Mother Rogers stopped to buy us a load of groceries. Janice had taken a week's vacation for her honeymoon. Monday morning, I was up early shaving before leaving on an out-of-town field trip with Dr. Dogger. Janice plopped herself playfully on the counter by the sink in her nightgown to watch me shave, and I said something impatient like "OK, honey, the honeymoon is over now."

Several months later and a few weeks after Christmas, Janice was in the hospital suffering from the emotional trauma of marriage to a man completely unfit and unprepared for marital responsibilities. Her doctor diagnosed her problem as "spastic stomach" caused by emotional stress and recommended that I not visit her for at least several days. Upon returning from the hospital to find the sink full of dirty dishes and the general mess in our apartment, she understandably didn't want to face it and asked me to take her to her mother who cared for her for a while.

By now, I had begun to think seriously about continuing graduate study toward the PhD. I would have another year of the GI Bill after completing the MS degree. Janice was working, and I would have an assistantship as well. She was not opposed to the idea but wanted me to stay at North Carolina State and close to home. I felt that broadening my horizons by studying elsewhere would make more sense. My thesis research would be complete in a few months. I was making top grades and being elected to recognized honor societies. Dr. John Lilly, an outstanding applied entomologist at Iowa State University, offered me an assistantship, of approximately three-year duration, to work with him on soil insect projects while pursuing PhD studies and research. I accepted the offer in spite of my wife's disappointment and that of her family's.

On June 6, 1954, Janice watched as I was awarded the MS degree in entomology on completion of research, thesis, and forty-five semester hours of credit with a grade point average of 4.0. Wives were also recognized with awards of the Good Wife's Diploma that Janice received and certainly had earned. My thesis was subsequently published in the *Journal of the Elisha Mitchell Scientific Society* 71(1):1-8, 1955, under title of Seasonal History Studies on the Southern Corn Rootworm and Its Relation to Peanut Culture in North Carolina.

Iowa State University. A few days later, I was on the Iowa State University campus at Ames, Iowa, and registered in the graduate school for the summer quarter. Janice had wanted to keep our car and rent a trailer to move our few belongings. I had stubbornly insisted that we could do without the car at first and save the cost of renting a trailer since married student housing apartments were already furnished with the necessities. So, I sold our car and took a bus to Iowa while Janice moved back in with her parents to continue her job with the Department of Motor Vehicles for the summer.

Most of my summer was spent digging in experimental plots of corn to count and record numbers of wireworms and corn rootworms. Dr. Lilly had three graduate student assistants until I arrived, which made four. One was writing his PhD dissertation while I was just beginning residence as a PhD candidate. Vern

Anderson and Roger Didricksen were pursuing MS degrees. The four of us shared an office next to Dr. Lilly on the first floor of the insectary building. From my arrival until late September, Roger, Vern, and I would be working in field experimental plots, in at least a dozen locations scattered across the state of Iowa, for several days each week collecting data on insect infestation and measuring crop growth or yields for different experimental seed or soil treatments. The three of us, with Dr. Lilly, put many miles on a state automobile and ran up sizeable travel accounts for hotels and meals on the road—all covered by money from research grants that our major professor held from a number of insecticide manufacturers.

In early September 1954, Janice rode the bus to Iowa, and we set up housekeeping on campus in a Pammel Court apartment for married students. I insisted that we could live without a car for a while at least. And so we walked everywhere we went, including to the grocery store almost a mile away. We celebrated our first anniversary by walking that evening to the movie theatre. The fact that neighbors all had vehicles didn't make matters any easier. And the fact that I often went to work and attended classes in denim shirts with GI-issue pants and a World War II Ike Jacket did not bother me as much as it did Janice. So, Janice got a job and went to work almost immediately with the Donnely Corporation in Nevada, Iowa, thirty minutes away. She got up daily before sunrise to catch her ride with a co-worker and to type addresses on envelopes all day before returning home after dark to cook and wash clothes on a washboard in the kitchen sink. With some of her money, we bought our first TV set with rabbit-ear antennas enabling us to receive two channels in black and white on an eighteen-inch screen.

While Janice was working to help buy bread, plus cooking, washing, and keeping house, I was beginning to work on my dissertation research, plus helping with Dr. Lilly's other projects while carrying eleven hours of class work, including statistical methods and a challenging chemistry course in radiotracer methods. The latter was a necessary preamble for some planned experiments in my research. We were both under much pressure and with little time or money to do much more than survive. One evening, I lost my temper and intentionally threw a yellow bowl on the floor breaking it in many pieces. Heated words were exchanged, and I suggested that she leave if she didn't like it. In late October 1954, Janice was packed and ready when Mother Rogers and Aunt Chloe arrived to drive her back home to Raleigh. Back home again, she went to work immediately with the state Department of Revenue there.

Meanwhile, I moved off the Iowa State campus about half a mile and into a rented room in a private home where I paid to eat breakfast and dinner with the family. I walked to work daily wearing a knit cap, Ike jacket, and gloves in the coldest winter I had ever experienced. One morning, I realized as I walked that it was unusually cold and my nose began to throb. On arrival at the insectary, I learned that the outdoor temperature was twenty-eight degrees below zero and that the tip of my nose was shining like a new dime. Fortunately, this early frostbite caused no permanent damage but was quite painful for at least an hour. Assistantship duties, research, and course work were taking practically all my time now; but I was also frustrated by loneliness and feelings of guilt about my marriage. My parents advised me to forget about marriage and just concentrate on school. For me, that was not so easy. I started writing Janice and telling her how sorry I was and how I really wanted to make our marriage work.

In March 1955, with a few hundred dollars, borrowed from my father by promissory note at 6 percent interest, I bought a 1949 white four-door Chevrolet sedan and drove back to Raleigh to get my wife. I told her that my schedule would not allow me to spend the night in Raleigh. And so within two hours of my arrival, we departed for Iowa. I don't know how much Janice drove on the return trip, but I remember waking up once on the backseat as she was navigating a sharp curve in the mountains of West Virginia. There was no housing available for us on the campus or in Ames on such short notice, and I would not have wanted to pick something out without her seeing it first anyway. In the nearby small town of Nevada, we found and rented a second-floor apartment with outdoor stairs. Mr. and Mrs. Metzger owned the property and lived with their large family on the first floor. As Seventh Day Adventists, their church worship was on Saturday, so they hung out their laundry on Sunday. This made their clothesline convenient for us also as Janice could hang our clothes out to dry on Saturday, her day off. She got a job almost immediately as an operator with the Iowa Continental Telephone Company and began to walk several blocks to and from her daily work while I drove daily to Ames and the Iowa State Insectary.

During this period, Janice met Irene Brown on the job, and they became close friends. Irene and Bob had two small boys and lived in a house in the country, although Bob was a student at Iowa State. We were invited several times to their place for a weekend visit, during which Bob and I would hunt pheasants and usually bring back three or four birds for an afternoon's effort. Bob graduated after another year, and they moved to Oregon, but Janice and Irene communicated with each other for several years. At last report, Irene had written and published several children's books. Besides the several weekend visits with the Browns and a couple of sight-seeing drives to nearby towns, we were not much involved in entertainment of any sort. However, the graduate students got together annually for a spring softball game with cold beer in place of water to quench the athletes' thirst.

Birth of a Child. In May 1955, Janice realized that she was pregnant. We were still almost two years away from graduation.

My eligibility for a regular monthly check for the GI Bill had finally expired. And we were expecting a child the following January. The future was bleak, and survival in graduate school was a matter of concern. I considered quitting and looking for a factory job or anything that might end the poverty. We both decided that if we could just find a way to hang on, even if we had to go further in debt, there would be a pot of gold at the end of the rainbow. And so we got on the list to move back to student housing in Pammel Court where rent was cheaper and gave up the apartment in Nevada. We were fortunate to get an apartment there in June. We painted the interior completely with both working almost every spare minute that we were not working elsewhere. Our unit was on a corner lot, and I dug postholes and built a fence around the entire yard in anticipation of the arrival of our first child.

 This was the fall quarter of 1955, and I was registered for ten hours of class work plus a noncredit course in reading knowledge of German to prepare for passing one of two language exams required of all PhD candidates. I had just passed my exam in reading knowledge of Spanish. Also, I had been trying for weeks to prepare for the preliminary oral exam that is one of the more frightening hurdles for the PhD degree. On the morning of this exam, Janice kissed me as I went out the door and watched me throw up my breakfast in the yard before leaving. I passed the exam after four hours of questioning by the six members of my graduate committee. Janice now drove the car back and forth to work in Nevada while I walked to and from the insectary. However, telephone company policy required her to quit working after her sixth month of pregnancy. And so another important source of income was gone. I took out a student loan for five hundred dollars from the university, and Dr. Lilly came up with a part-time job for Janice counting daily moth collections from light traps until it got too cold for insects to come to the traps. That helped for a few weeks, and then we needed a second loan that was also granted.

Janice was at the end of her eighth month by Christmas, when Aunt Chloe came to visit for three or four days. I met Chloe at the bus station and showed her the entire town of Ames in about five minutes before going back to the campus and home. Chloe was entertained the next day with a campus tour and visit to the insectary building that housed the department of entomology. After that, it was too cold to do much else but sit by the heater and visit. Snow and ice were everywhere, and the roads were packed with it, a condition commonly prevailing there for about four months of each year.

Three weeks later, while Janice was running clothes through the wringer of our washing machine, her amniotic sac ruptured. I was called at the lab and rushed home to drive her to the hospital. The pediatric nurse examined her and said there was no rush and that it would be several hours at least before the baby arrived. I had started a series of tests in which observations were made at two-hour intervals and eight times daily for eight consecutive days after which the tests were repeated four more times for a total of forty days of observations. So I rushed back to the lab to record some scheduled observations. Upon my return to the hospital, Janice Faye Long had already arrived at 5:00 PM, January 20, 1956. When Janice awoke, I was standing beside her bed to congratulate her and admire our beautiful daughter Jan. Unfortunately, I had not been there when they had put Janice to sleep. She contracted an infection with fever that continued for almost a week before they let her go home. Mother flew from Alabama to help with baby Jan for several days. We appreciated Mother's help, and certainly, she did take much of the load off Janice for several days. However, her actions in the kitchen and with the baby made Janice very nervous. Mother returned home after four or five days with our sincere thanks.

during the winter months, they usually froze as hard as boards before being brought inside to dry while hanging from doors or ceiling.

In July 1956, Aunt Chloe sent money for Janice and the baby to fly to Raleigh for all the family to admire and for Janice to get a much-deserved break. After several days, I drove to North Carolina for a brief visit with everyone there before taking Janice and Jan on to Alabama to see the paternal grandparents. Daddy was much taken with the baby, who was his first grandchild, and enjoyed sitting in the backyard swing with her in his lap. I believe that Mother's intentions were good and that she really tried, but she was never able to conceal her negative feelings for Janice. Parting company was always a relief for us and perhaps for Mother as well.

I could not nurse the baby, but I enjoyed rocking with her in the rocker as often as possible. I did bring in the frozen items from the clothesline, grocery shopped, changed the baby's diapers when necessary from 11:00 PM to 6:00 AM, handing Janice the baby to nurse and putting the baby back in her crib afterward. At the same time, I was carrying eleven hours of class work each quarter and studying German from midnight until 2:00 AM nightly from January until May 1956 when I finally passed the exam. I conducted experiments continuously through October 1956, writing a two-hundred-thirty-one-page dissertation from October 1956 through February 1957, and defended it before my graduate committee the next month.

Job Selection. In late fall of 1956, it appeared that I might be completing all degree requirements by late winter 1957. There was a job opening in Hawaii for a research entomologist with clearance to work with radioisotopes. Although I was qualified, the idea of forever wearing a white lab coat while handling radioactive materials for the rest of my life did not appeal to me. Instead, I opted to interview for two university research positions, one at the Virginia Polytechnic Institute in Blacksburg and the other at Louisiana State University in Baton Rouge. And so in mid-December 1956, I flew from Iowa to Blacksburg to meet Dr. James Grayson, head of the entomology department, and from there to Baton Rouge to meet Dr. Dale Newsom, head of the department of entomology at LSU, before returning home. Dr. Grayson offered me a job with responsibility for peanut insect research at one of the Virginia experiment stations. I had a similar offer from Dr. Newsom to assume responsibility for sugar cane insect research, but with office and laboratory on the main campus in Baton Rouge. I had left Iowa covered with ice and snow to find more of the same in Virginia. The night before my interview with Dr. Newsom, I stepped out of Pleasant Hall on the LSU campus for refreshments and was greeted by green banana trees and blooming poinsettias. I hastened to call Janice to tell her that I thought we had found a home in Louisiana. She liked my description of the climate and scenery but was disappointed to hear that I wanted to pass up a job only a couple of hours' drive from her home in North Carolina.

The annual meeting of the Entomological Society of America was held in New York City. Dr. Lilly had insisted that I attend the meeting to report my research findings, an idea about which I was not at all enthusiastic. However, with a neatly typed eight-minute presentation and with good color slides to accompany my talk, I flew to New York and presented a paper on the behavior of wireworms in response to different insecticide treatments on corn seed. Roger Didriksen was visiting family that week in White Plains, New York, and insisted that I go with him to the New Year's Eve celebration at Times Square. While we stood for several hours elbow to elbow in a crowd waiting for the big apple to fall, I needed to pee and felt close to frostbite. It was an exciting event for Roger but a miserable one for me. I called Janice, who was at home with the baby, in hope of cheering her up but learned later that it had only added to her depression. She now reminds me that, during her many months in Iowa, the only places she saw besides Ames and Nevada, where we had lived there, were Pella and Des Moines. One Sunday, we drove to Pella, a small town with a Dutch flavor, to see the tulips in full bloom, and I took her to Des Moines on another occasion to buy a maternity dress. Such were the times.

Two years and nine months after completing the MS degree, I had finally completed requirements for the PhD in entomology with minors in soil science and ecology. Included in the process were one hundred eleven quarter hours of credit beyond the MS with a grade point average of 3.84. Words cannot describe our sense of accomplishment and relief in the knowledge that we had survived an endurance test the likes of which we hoped to never see again. My degree was awarded in absentia at commencement exercises on June 15, 1957, by which time we had been living in Baton Rouge, Louisiana, for three months.

9

LSU Faculty Period (1957-65)

In late March 1957, we departed Iowa with our fourteen-month-old daughter and a loaded U-Haul trailer containing all our belongings, except for those in the Chevy with its four slick tires. We left behind us ice, snow, and brown vegetation and headed south for a new life and new experiences. As we moved through Missouri, Arkansas, and into Louisiana, we saw increasing amounts of green vegetation that further lifted our spirits. A graduate student friend, Dr. Ed Burns, who had preceded us in accepting a job at LSU, had located a temporary rent house for us. So, we just drove into town and parked in front of a two-story white frame house with tall colonial columns and a narrow front porch. The house appeared to be relatively new in an older blue-collar neighborhood. Upon entering the front door, we immediately encountered stairs that led to two bedrooms and a bath. Downstairs there was a living room to the left of the staircase and small dining area and kitchen to the right; and that was it. The house was really small and nowhere near the size that it appeared to be from the front.

A few days later, our next-door neighbors, Mike and Janelle DeLaune with their five children, boiled crayfish in their backyard. We had never eaten crayfish before and thoroughly enjoyed doing so. Janelle and Mike were very friendly and enjoyed educating their new redneck neighbors to the facts of south Louisiana culture. We learned right away that all natives of south Louisiana were either cajuns or coonasses. Cajuns were descendants of immigrant Canadians and coonasses were all the rest. If you were a newcomer who lived here for a while and fitted in well with the culture and was well-liked, you might be privileged one day to receive a certificate of adoption as a "bonafide coonass." However, I soon learned that this introduction by the DeLaunes was somewhat oversimplified and that south Louisiana, beginning with the New Orleans area, was a tremendous melting pot of people and their descendants from many parts of the world.

We no longer benefited from the partial furnishing of a Pammel Court apartment. And so we desperately needed a dining table with chairs as well as a washer and dryer. During the trip from Iowa, my iron typewriter had fallen through the upholstery and into the back of the old couch we had brought with us, so we also needed a couch and another piece or two of furniture for the living room. I don't know how we survived until the first paycheck, but I do remember that my first monthly check, less eighty-two dollars in deductions, amounted to a net total of three hundred fifty-two dollars. With rent more than doubled now at seventy-five dollars per month, we realized that we had not yet reached the Promised Land. Dr. Newsom cosigned a note allowing us to purchase some much-needed household furnishings.

The *Sugar Bulletin*, published monthly by the American Sugar Cane League, ran an article in its June 1, 1957, edition with a photograph of Dr. William Henry Long, announcing his recent employment by the Louisiana Agricultural Experiment Station as assistant entomologist to conduct sugar cane insect research. The article briefly reviewed my work in North Carolina and Iowa and mentioned my special interest in soil insects. Dr. Newsom was quoted as saying that "Dr. Long is the best qualified young man in the country to tackle the problem of soil insects in sugar cane."

The Challenge. There was certainly a need for research on soil insects. However, there was an even more pressing need for studies to find a solution to the heavy crop losses then being suffered annually from damage by a stalk-boring caterpillar known as the sugarcane borer (SCB). During the next few years, my students and I, with the help of a few others, tackled these problems with such success that we significantly changed a number of popular agricultural practices over a period of half-a-dozen years. These changes were very beneficial to the sugar industry; however, there is always opposition to change. And the greater the change, the greater the opposition. These were no exceptions to that rule. Businesses were lost and new ones begun. After initial resistance from most directions, many converts were made; some became friends, others only reluctant admirers. Some became lifelong enemies.

My predecessor in the sugar cane research job had been Mr. Al Dugas. Mr. Dugas had earned his MS degree from Iowa State some twenty years at least before my time and had been a member of the LSU entomology department until two years before my arrival. He was a bonafide coonass aristocrat, loved and respected by movers and shakers in the sugar industry as also by a number of LSU administrators and state legislators. Mr. Dugas had resigned from LSU when he was passed over for the department head's job in favor of Dr. Newsom. Al Dugas had numerous friends in high places; but when push came to shove, Dale Newsom was in all company by far the best informed, most articulate, hardest working, most respected, best leader, and all-around best qualified agricultural entomologist in sight. Fortunately for LSU at that time, retired Maj. Gen. Troy Middleton, commander of the Eighth Army Corps in Europe during World War II, was president of the university. Under his command, politics counted little. He was interested primarily in facts, let the chips fall where they may.

Prior to my arrival, Dr. Newsom had initiated field studies with endrin insecticide showing over 90 percent control of the SCB with large associated increases in crop yields. The insecticides being recommended by the LSU extension service and used by growers at the time gave only about 50 percent control even when used season long on fixed schedules of weekly applications. I continued the endrin studies started by Dr. Newsom and also began systematic sampling and dissection of plants throughout the year to identify and record observed life stages of the SCB and to determine where on or in the plant each stage occurred. Mr. Dugas was currently mass rearing and selling *Trichogramma* wasp parasites of SCB eggs to farmers on the assumption that these would help control those SCBs not being adequately controlled by the relatively ineffective insecticides then being used. Therefore, I began collecting SCB egg masses from plantation fields where parasites were being released and from other comparable fields in which they were not released to evaluate the effectiveness of the parasites. Similar collections also were made from endrin-treated and untreated field plots. These egg masses were held in the laboratory to determine average amounts of egg parasitism occurring in the different fields. I had the benefit of a full-time assistant technician, Mr. Emile "Max" Concienne, who had been Mr. Dugas's assistant for years. After a few months, a graduate student from India, Kamta Katiyar, was assigned to my project as a half-time graduate assistant. Katiyar began to rear SCBs in the lab under controlled conditions and to determine the times required for each developmental stage. These additional hands and Max's experience were a tremendous help in the tedious collection of a great deal of useful information.

One early morning in June 1957, on a landing strip at Oaklawn Plantation, I was introduced by Max to Mr. Scott Tibbs, an aerial applicator who dusted a lot of sugar cane in St. Mary and Iberia parishes. He had often cooperated in the past with Max in treating experimental plots for Mr. Dugas and was ready now to work with me as Max handed him a map of the plot layout. Mr. Tibbs took a deep breath and exhaled slowly as he looked at a sheet of paper with fifty-four squares in nine different colors, and then looked out across the field in which colored flags marked the experimental plots for nine different treatments, each replicated six times. Individual plots were approximately six hundred feet long by one hundred fifty feet wide that would require him to take off with less than half of a normal load and fly at nearly a hundred miles per hour while distributing each load equally over six different plots. Suddenly, he exploded angrily with the question "What stupid ass designed this experiment? Dugas never did anything this dumb!" I thought that for starters, that's a good reason to change but responded that "from a statistical standpoint, we needed to decrease experimental error by randomizing treatment assignments to plots and by increasing replication." I don't think that this made much sense to Mr. Tibbs, but he did the job and repeated it three more times at two-week intervals.

In 1957, the U.S. Department of Agriculture, in cooperation with the state departments of agriculture, embarked on an ambitious program to eradicate the imported fire ant by broadcasting heptachlor insecticide over every square foot of millions of infested acres in southeastern states from Texas to Florida. In Louisiana, aerial applications began in early spring in central and northern parishes. However, by early summer, over two thousand acres of sugar cane in several southern parishes also had been treated. By midsummer, devastating crop damage of a high magnitude never before seen from the SCB was showing up in heptachlor-treated fields. I hastened to set pit traps in several dozens of these fields with equal numbers in nearby untreated fields for comparison. We found that ants were by far the most abundant arthropod predators of insects present in the untreated fields, that they were practically absent from the heptachlor-treated areas, and that there was a high correlation between the absence of fire ants and crop damage by SCBs. Further field observations revealed fire ants attacking SCB larvae of various sizes. The results of this survey were published in the *Sugar Bulletin* 37(5):62-63 under the title of Fire Ant Eradication Program Increases Damage by the Sugarcane Borer. Rachel Carson in her explosive best-seller book of 1962, *Silent Spring*, cited this article in support of her statements to the effect that heptachlor applied to Louisiana farmlands unleashed one of the worst enemies of the sugar cane crop. She reported that the chemical aimed at the fire ant had killed off the enemies of the SCB and that the crop was so severely damaged that farmers sought to bring suit against the state for negligence in not warning them that this might happen.

Further studies by me and co-workers documented the fact that ants, particularly the red imported fire ant, were of major importance in the natural control of the SCB and led us to initiate a policy of conservation of these arthropod predators by avoiding the use in cane fields of insecticides that were very detrimental to ants. Subsequently, my correspondence with Senator Ellender, longtime senator from Louisiana, helped him to convince the U. S. Congress to discontinue federal efforts to eradicate the imported fire ant from the United States.

As endrin insecticide continued to provide outstanding SCB control in our field tests, rumors began to circulate that a change in control recommendations might be coming soon. Dugas had long opposed use in sugar cane of the newer, more effective insecticides on grounds that they would destroy beneficial insects like *Trichogramma*, which he sold annually at a price of fifty cents per acre to farmers growing several hundred thousand acres of sugar cane. LSU and USDA researchers, studying the efficacy of artificial releases of *Trichogramma* for SCB control in Louisiana, had never reached agreement. USDA studies had always suggested that such releases had no measurable impact on SCB populations. On the other hand, published LSU studies, including a recent one by Dugas, showed very economical and beneficial results. Besides selling parasites, Dugas sold a scouting service, for one dollar per acre to the majority of cane growers in the state, by which farmers were advised of the proper time to begin insecticide operations. The official LSU SCB—control recommendations, which he had authored and were still in effect at the time, were irrational, relatively ineffective, and impractical in ways that caused Dale Newsom to refer to them as "Dugas's voodoo." The mystery of the whole system prompted growers to appreciate their need for assistance from their consultant, Al Dugas, who was the only independent sugar cane entomology consultant in the state at that time.

For obvious reasons, Mr. Dugas became much concerned about the possibility of any changes in recommendations for SCB control. When he came to see Dr. Newsom to find out what was going on, Dale referred him to me. I showed him the continuing outstanding performance data for endrin insecticide and suggested that we would probably recommend it to farmers next year (1958). Although he was nearly thirty years my senior, he did try to respect my position while seeking to enlighten me on the subject of long-held practices and theories about SCB biology and control. The more we talked, the more apparent it became that there was very little that we could agree upon. After about an hour, it was plain that his blood pressure was rising as was his voice and his face was becoming increasingly flushed. Before completely losing his cool, he left my office to go back to Dr. Newsom. There in Dale Newsom's office, he told Dale that these contemplated changes in recommendations would not stand and that he would not let them stand, before storming out and slamming the door. Dale's response, heard by a few up and down the hall, had suggested that he not darken his doorway again.

Within the next few days, Dugas visited Dr. Charles Upp, director of the agricultural experiment station, by whom he was referred to Dr. Norman Efferson, dean of the College of Agriculture. Neither of these gentlemen was prone to bite this bullet, and neither would have wanted to be on Dr. Newsom's "shit list," although both were above him in the administrative hierarchy. So then failing to get satisfaction elsewhere, Mr. Dugas went to the office of the president of the university. General Middleton listened politely before telling Dugas that the decision was in the hands of LSU entomologists and that their decision would be final. In response, Dugas threatened Middleton with a promise that changing the SCB recommendations would result in financial punishment of the university by the state legislature. General Middleton picked up his phone and called Dale Newsom. "Dale, do you, folks, know what you're doing with those SCB recommendations?" The reply was "Yes, General Middleton, we do!" The general said, "Then have at it, Dale." Dugas left the general's office determined to start a war.

All project leaders in the entomology department were required to write an annual report of their research accomplished during each calendar year. In the case of the sugar cane project, the report was written also for publication and distribution to the contact committee of the American Sugar Cane League. I soon learned that this committee meeting was an annual affair attended by several hundred league members, including both factory and field personnel. It was held in a campus auditorium and usually was an all-day affair during which each project leader had his report distributed to the crowd before discussing and summarizing its contents from a podium on center stage. Dr. Newsom advised me in January 1958 that I should be preparing for that late March event. I began to realize that even entomologists cannot forever escape the spotlight of public attention, and that I must either find a way to stand up and speak out on matters of responsibility, in spite of my overwhelming fear of public speaking, or I would be relegated to the trash heap.

    Dugas and his supporters began to spread a gospel of fear that the Louisiana sugar industry was about to be led astray by entomologists without much experience with sugar cane, some of who were "just out of school and still wet behind the ears." He pictured the destruction of the industry by new practices that would eliminate the benefits of parasites and predators of pests. His influence was such that most industry leaders as well as scientists from other disciplines on the LSU campus were inclined to support his position. Storm clouds began to gather as increasing numbers of people voiced their concerns and opinions. The approaching contact committee meeting was shaping up to be a shootout at which the young and inexperienced Dr. Long would be thoroughly roasted. I was even warned by one farmer to stay out of his fields or be prepared to be thrown out. I was referred to by some as "that guy who's trying to destroy the sugar industry."

    On the day of the big meeting in late March 1958, Mr. Concienne arrived at my office at 9:00 AM to pick up stacks of our report and take them to the auditorium. As I wasn't scheduled to speak until 11:00 AM, I told him to go ahead and that I would come along later. I sat alone for another hour and forty minutes going over notes and doing a last-minute check of details and data that might be helpful in arguing my points. At precisely 10:40 AM, I opened a thermos bottle and drank six martinis, without pausing, before closing my office door and walking eight minutes to the auditorium. At the podium, I pointed to outstanding results from three years of research (two of Newsom's and one of my own), showing previously unheard-of levels of SCB control by endrin insecticide accompanied by similar increases in crop yields. I emphasized that these results were obtained not from eight to twelve or more weekly insecticide applications but from only four made at fourteen-day intervals. When asked by Mr. Dugas about the harmful effects of these treatments on beneficial *Trichogramma* parasites of SCB eggs, I replied that there were no such effects and supported my statement by showing my own data on levels of SCB egg parasitism from endrin-treated and comparable untreated fields. I added that even in the absence of insecticide treatments, egg parasitism by late summer was similar in both *Trichogramma*-released fields and those in which parasite releases had not been made.

I then proceeded to attack his methods for determining when and where to treat for SCBs by pointing out that numbers of dead plant shoots in the spring often had little to do with significant numbers of SCBs present and the need for insecticide treatment, since, by that time, many SCBs either were buried deep within the plant, where the insecticide could not touch them, or they had matured and flown away. I also stated that the counting of SCB egg masses, as he recommended, was so difficult and time-consuming, even for a trained entomologist, that it was useless for farmers. I maintained that looking for small larvae in or behind plant leaf sheaths was easier and more pertinent to the problem as these were the ones susceptible to insecticide exposure while larger larvae were mostly inside stalks into which they bored about ten days after hatching. There were questions from pathologists, agronomists, and others who voiced concerns to which I believe I responded appropriately. Although the enemy had been wounded, he and his army were by no means yet defeated. There were many battles still to come in meetings with county agents and farmers in the different parishes as well as future contact committee meetings.

The six martinis had saved the day. I felt that I had never before experienced such success in verbally vanquishing an enemy. However, the fact that I had depended on alcohol to do it

Six days after leaving the hospital, as she lay on our sofa while friend Lois Didriksen did some ironing, Janice began to hemorrhage. Lois called me at the lab, and I rushed home to find her losing a frightening amount of blood into a dishpan. We wrapped baby Jan in blankets and packed Janice with towels to soak up the blood as I drove them both back to the hospital. It was late afternoon when we arrived and early evening when Dr. Fellows finally saw her and recognized that she was in shock. There were no resident interns or even a blood bank at this little country hospital in 1956. I gave a pint of blood by direct transfusion and then trudged through the snow from door to door at Pammel Court asking for volunteer blood donors. Several graduate student friends volunteered, and a few students I didn't know at all donated blood. Janice had D&C surgery that evening and received several more pints of blood before returning home several days later. Mother Rogers flew up while Janice was in the hospital to help for a week before returning home. She cleaned, cooked, washed, and cared for the baby, all of which helped immensely.

At this point in February 1956, we still had thirteen months to go before graduation with regular income of only two hundred dollars monthly from my assistantship. University policy would not permit the school to lend me any more money. However, Dr. Lilly arranged for me to make a few hundred extra dollars editing catalogs for two distributors of agricultural chemicals by adding state-recommended uses for their product lines and arranging all pertinent information in easily accessible tables for publication in booklet form. Chloe was sending us fifteen or twenty dollars monthly at this time, and our parents responded when called upon from time to time for forty or fifty dollars, all without strings attached. Janice was nursing the baby, which she did for eight months, putting food on the table and keeping house. Clothes were hung year-round on a line in the backyard to dry. However, really began to bother me. I could imagine myself becoming completely dependent upon it in this job. Yet the fact that I could perform so well at all was proof I thought that somewhere within me was the ability to do so if I could just find the right buttons to push. In six months, I would be thirty years old. I had been studying Mormon (LDS) church doctrine for several years and was seriously considering joining my wife in her faith. Requirements for baptism, in addition to the usual requirements of faith and repentance, included abstinence from alcohol and tobacco. I had been a slave to both for thirteen years. And so at about this time, I determined once again, after many previous attempts, to give them both up completely. It was very difficult, but after six prayerful months, I no longer craved nicotine. I didn't crave alcohol as if my body would perish without it, but I very much missed its socializing effects. And I could be a real basket case in social situations without it. Nevertheless, I found that I could live without it and did so for the next fifteen years.

 A new PhD entomologist and graduate from Oklahoma State University, Dr. Sess Hensley, had accepted a job with the USDA sugar cane laboratory at Houma, Louisiana, in December 1957. His boss and senior entomologist there, Mr. Ralph Mathes, was not comfortable with the current conflict over SCB control practices. Mr. Mathes, like so many politicians, tended to go with the flow and was uncomfortable in taking almost any stand and particularly one at this time with LSU entomologists engaged in the current controversy. Sess, on the other hand, appreciated what we were doing at LSU and was anxious to join with us in cooperative research. He had a full-time assistant, Mr. McCormick. And so our research team increased by two more pairs of eyes and hands. In spite of continuing objections from his boss, Sess and I began to plan our project work jointly, sharing results and often jointly publishing them. Sess was a fortunate Eighty-second Airborne veteran and survivor of World War II and the Korean War. He liked to mix and mingle with folks and did a lot of it in the course of our work with farmers, plantation managers, and sugar mill operators. I liked to work but was not much at mixing and mingling. We regularly compared notes and benefited mutually from our respective observations.

The *Franklin Banner-Tribune* for February 18, 1958, reported in detail, with a picture of the participants, a meeting of the Sugar Producers' Club held in New Iberia and attended by prominent sugar industry leaders in addition to Dr. Newsom, Mr. Dugas, Mr. Ralph Mathes, Dr. Hensley, and Dr. Long. I presented all available research results on endrin for SCB control, stating that we had observed no negative effects of the insecticide on *Trichogramma* egg parasites and that there was no consensus of research to show that releases of artificially reared egg parasites were beneficial. Mr. Dugas replied that the endrin data was limited, that he could not understand why ryania insecticide was left out of the recommendations, and that he didn't understand why anyone should be confused on the use of parasites. He considered them a supplement to insecticides and doubted if cane could be grown in Louisiana without the parasites. Newsom disputed a number of Dugas's statements and said that there was disagreement around the world on the use of *Trichogramma* parasites. And so the meeting ended with Mr. Dugas's cronies wondering who they should believe.

In the spring of 1958, I flew to Honolulu, Hawaii, to present a paper to the International Society of Sugar Cane Technologists on the subject of recent SCB control studies in Louisiana. I presented my paper, answered questions afterward, and sat down. I was the youngest entomologist there and the only one who believed that insecticides should have any role in controlling moth borers in sugar cane. Practically, all of these men were from more tropical climates in which biological control measures generally are more effective. Also, they had been longtime acquaintances or friends of Mr. Dugas and were predisposed to view anything I might have to say with skepticism. After the meeting sessions, many gathered in the bar for drinks and conversation. At that time, I didn't drink and had never been much at idle conversation either. There was such opposition to what I was doing in Louisiana that almost any conversation was forced. Except for touring the island of Oahu with an earlier graduate student friend from Iowa State and now faculty member at the University of Hawaii, I spent the major part of two days sitting in meeting sessions. I tried unsuccessfully to ride a surfboard early one morning and brought home a field-ripened pineapple for Janice and Jan.

Strength from Faith. Prior to our marriage, I had promised my fiancée that I would earnestly study Mormon church doctrine by reading all the standard works of the Latter-day Saint Church, by studying other books on the subject and by attending meetings with her until I could make a decision with which I was comfortable. I lived up to my promise, and on September 7, 1958, in Baton Rouge, Louisiana, I was baptized by immersion (after infant baptism in the church of my parents) a member of the Church of Jesus Christ of Latter-day Saints after all that my mother could say or do to prevent it, which is another story of its own. While a newly appointed professor at LSU, I was immediately assigned by the ward bishop to join the cleanup crew on weekends to keep the chapel and grounds in order. A few weeks later, I was asked to speak briefly (five minutes or less) during sacrament meeting. Subsequently, I was ordained a deacon and made an assistant ward clerk, then a teacher (February 1, 1959) and assigned a Sunday school class of about fifteen teenagers. Later (May 3, 1959), I was ordained a priest and given responsibilities in the ward clerk's office.

Then I was assigned for the first time to deliver the main talk (sermon) at sacrament meeting. The date was September 20, 1959, my thirty-first birthday. I was a research professor at that time, without university teaching responsibilities and with little public-speaking experience. The testimony that I had was still fragile, and the thought of trying to influence for good so many people by my own voice was frightening. I was seated on the speakers' platform in the Baton Rouge ward chapel with sweaty, trembling hands and a house full of people as I heard my name called in announcement of the program. I began to pray earnestly: "Please, God! Help me to be a useful tool in your hands this day and to say something beneficial to these people." Within a few seconds and as if by supernatural intervention, which I believe it may have been, I experienced almost total relief from anxiety. When introduced a few moments later, I rose with confidence and began to speak. What I said was not particularly profound, but for me, it was a great step forward both in self-esteem and in faith.

While in the Baton Rouge ward, I was ordained an elder September 18, 1960, and was kept busy teaching in the Sunday school and assisting with ward clerk duties. Another priesthood assignment given me at this time was to establish a branch Sunday school in nearby Denham Springs and to serve there as the first Sunday school superintendent. This Sunday school later became a branch of the church and then a ward. There were other priesthood duties and responsibilities during this period, but I do not remember them all. My wife reminds me that one was a call to serve briefly as stake Sunday school superintendent in the New Orleans stake.

During the second quarter of 1958 and beyond, Sess and I both attended numerous farmer meetings with county agents in the various sugar-producing parishes. SCB recommendations were always a major concern at these times. At some meetings, we were both present. At many others, either he or I would represent entomology on a program that also often included several other specialists, such as an agronomist, a pathologist, a plant breeder, or an economist. One very memorable meeting was held in Lafourche Parish in the second-floor auditorium of the agricultural building at Thibodaux, Louisiana. Sess and I were the principal presenters in entomology at the meeting. But it was also attended by our bosses—Mr. Mathes and Dr. Dale Newsom; by Dr. Newsom's bosses, Drs. Upp and Efferson; and by several LSU department heads in agriculture, including the head of the plant pathology department, Dr. St. John P. Chilton. Chilton and Dugas were longtime friends and drinking buddies. Chilton, as a sugar cane pathologist, had been accustomed to sharing with Dugas the ruling of the roost among sugar cane scientists, and tended to view Newsom, who had cut his teeth working on cotton and soybean insects, as an outsider to the sugar cane fraternity.

After my brief presentation, followed by supporting comments from Sess, Dr. Upp stood to remark that with all due respect to these well-trained young scientists, Long and Hensley, perhaps the industry should go slow in adopting these radical changes and not be too hasty. Dr. Chilton then rose to attack the validity of our data or its interpretation. Before Chilton could yield the floor, Dale jumped to his feet to eloquently chastise Chilton for dabbling in areas about which he knew little and suggested that he confine his attention to the few things that he did know something about. With that, the county agent announced that it was time for a lunch break. As the crowd filed out of the room and down the stairs, Chilton was overheard mumbling his objections to something said, whereupon Dale spoke in a voice loud enough for all to hear, "If that son of a bitch opens his mouth again, I'll shove my fist down his throat up to my elbow." The meeting continued after lunch on subjects other than entomology, and Dr. Chilton made no more public statements.

Recognizing the strength of our opposition, but also believing that nothing was stronger than the truth, we planned ten demonstration experiments in as many different parishes—all easily visible to passersby from heavily traveled roads. "And ye shall know the truth, and the truth shall make you free" (John 8:32). At each location, three large side-by-side plots of sugar cane were identified by large signs as RYANIA, UNTREATED CHECK, and ENDRIN, respectively. The ryania- and endrin-treated plots were treated on the same dates by aircraft according to the new endrin recommendations for 1958. Treatments were begun in all treated plots by mid- to late June. By mid-August, all insecticide-treated plots had received four applications at fourteen-day intervals, and all endrin-treated plots stood out clearly greener and taller than the ryania-treated and untreated check plots of obviously shorter cane with numerous brown and broken tops. These sights quickly became common knowledge and were much talked about throughout the industry. Relatively few farmers had followed the new recommendations in that year of 1958.

Contributions and Recognition. However, the next year, practically everyone was using the cheap and effective endrin, whether they needed it or not. On March 26, 1959, I was awarded a friendship bond of $1,000,000 in "good wishes" from the Louisiana National Bank, signed by the bank president and recorded in the Baton Rouge Morning Advocate. On April 6, 1959, an article in the Farm News section of the *Lafayette Advertiser* entitled "Scientist Finds Control for Sugarcane Borer" pictured me at work in my office and stated that "sugar yield increase was about twice as much as from crops treated with ryania dust, the most effective chemical used prior to endrin granules." The *Franklin Banner-Tribune* for December 24, 1959, ran a front-page article entitled "St. Mary sugar men enjoy banner year," in which the treating of crops with endrin for SCB control was cited as an important factor in the success of that year's crop. The St. Mary county agent estimated that for every dollar invested in endrin, farmers received $4.11 in return and that net returns per acre were approximately $25.00 for 31,066 acres treated with endrin in the parish that year.

By 1960, we were able to reduce overall endrin use about 25 percent by increasing the interval between applications from fourteen to twenty-one days, thus reducing the total number of applications from four to three without losing any of the benefits from treatments. Mr. Dugas's sale of egg parasites had dropped precipitously until he finally closed down the *Trichogramma* operation that year. Now, with so many acres being treated with endrin, regardless of need, a problem arose with insecticide fish kills from field runoff following heavy rains. We had known from the beginning that endrin was quite toxic to fish and had anticipated some fish kills, but nothing of the magnitude we now encountered when heavy rains followed widespread endrin treatments. We had believed that the need far outweighed the risks, and many people in the south Louisiana area of a sugar-based economy agreed. However, farmers would also admit that rabbit and quail populations were not what they had been three years earlier. As biologists, we believed that if it were possible to swat enough flies with a flyswatter, one could eventually produce a generation of flies completely resistant to flyswatters. Therefore, we had already begun to look for alternative insecticides and safe procedures for further reducing insecticide use without giving up the benefits of good SCB control.

In 1962, the LSU Foundation proposed a need for $524,055 to support twenty-three projects of which number seven was to "prepare and publish a book on the biology and control of sugar cane insects in the world by Dr. W. H. Long, director." It was stated that "such a book would provide the sugar cane industry, researchers, and students with complete up-to-date information on insects which affect sugar cane." I was flattered of course, but had no time then for such an activity, and honestly hoped that it would be forgotten for at least a while.

In 1964, I wrote and published the LSU recommendations for sugar cane insect control (*Sug. Bull.* 42(15):183-84), which offered a new, effective, and alternative insecticide, guthion, in the event of resistance to endrin, and which described for the first time a practical weekly field-scouting technique and a SCB population threshold for treatment based realistically upon the presence of small larvae on 5 percent or more of stalks examined. This was designed to end all insecticide treatments by scheduled intervals since new pest population assessments should now be made before each insecticide application. In other words, fields should be scouted weekly and treated only when borer infestations exceeded that 5 percent threshold.

These were timely recommendations, as the first instances of SCB resistance to endrin had occurred in August of the previous year, 1963, after only five years of heavy endrin use. I went to Florida to collect and bring back SCBs never before exposed to endrin. By rearing these in the laboratory and comparing the susceptibility of their progeny to endrin with that of similarly reared Louisiana SCBs, we found that Louisiana SCBs could then tolerate more than a hundred times the amount of endrin required to kill Florida SCBs (*J. Econ. Entomol.* 58(6):1122-24).

Prior to 1958, insecticide applications for SCB control had been recommended earlier and might continue, with recommended groups of four weekly applications for a total of eight to a dozen times during one growing season. It had been generally accepted that at least three annual SCB generations were potentially destructive to the crop (*Sugar Bulletin* 34(13):191-2). We published studies in 1963 (*J. Econ. Entomol.* 56(3):407-9) showing that controlling first generation SCBs was not economical and that the sugar cane plant compensated adequately for first-generation borer attack by producing more stems or tillers than could possibly reach maturity. In 1964, I published additional studies (*J. Econ. Entomol.* 57(3):350-53) showing that the most critical period for controlling the pest begins with the first appearance of internodes on the cane stalks, that this commonly coincides with the beginning of the second SCB generation, and that susceptibility to damage gradually decreases thereafter until there is no justification for insecticide use after early September. These studies resulted in reducing the needed number of insecticide treatments from as many as twelve to three or less per year with less costs to the farmer and far better insect control.

I wrote the new 1964 insect control recommendations (*Sug. Bull.* 42(15):183-4), which incorporated the above information and also discontinued the general use of chlordane insecticide in the planting furrow of all sugar cane fields of all soil types. This had been practiced at significant costs for more than a dozen years without sufficient justification. I had found no significant yield differences between chlordane-treated and untreated plots from 1961 through 1964 in thirty-nine locations scattered about the sugar cane-growing area of the state (*J. Econ. Entomol.* 60(3):623-29). Extensive examination of soil samples from treated and untreated plots over a two-year period showed no significant differences in populations of potentially damaging insects or other arthropods. We demonstrated that soil treatment with chlordane, in the absence of any potential pests, had a stimulating hormonal effect on early cane growth, but was an effect that did not translate into better crop yields. The fact that chlordane was a very effective ant poison also rendered it undesirable for use in cane fields where we had shown ants to be highly beneficial. For these reasons, chlordane was recommended for use only in sandy soils and previously grassy fields where wireworms were a likely problem.

These 1964 recommendations marked the beginning of a new era in sugar cane entomology in Louisiana and began the earliest general practice of integrated pest management (IPM) in a major field crop in the southern United States. Such regular field scouting and reductions in insecticide use would (1) prolong the useful life of new insecticides by delaying the development of resistance, (2) help avoid adverse effects of insecticide residues on crop growth or on beneficial ant predators, (3) reduce fish kills and adverse effects of insecticide on other wildlife, (4) reduce hazards of insecticide contamination of drinking water, (5) eliminate the unnecessary expense of applying insecticides when they are not needed, and (6) permit greater realization of benefits from the inherent resistance to insects present in certain sugar cane varieties.

While research entomologists had written, published, and assumed most responsibility for sugar cane insect control recommendations during the '50s and '60s and through 1964, now other members of the LSU community suddenly desired to be identified as visible members of a winning team. The 1965 recommendations (*Sug. Bull.* 43(17):204-05), which differed in no important way from those of the previous year, were published with John A. Cox, director of the LSU Division of Agricultural Extension, as sole author. Dr. Cox was not an entomologist and had extension entomologists under his command, but now he obviously desired to let it be known that the Division of Agricultural Extension wholeheartedly supported these recommendations and, henceforth, would assume responsibility for their dissemination.

A significant amount of information obtained by my project during these years resulted from the efforts of graduate students pursuing thesis or dissertation research toward their MS or PhD degrees. During eight years on the LSU faculty, I had served as chairman of ten graduate student committees, for five MS and five PhD degree candidates. Important parts of the chlordane studies were parts of Howard L. Anderson's MS thesis and Abdel Latif Isa's MS thesis and PhD dissertation. The laboratory studies of endrin resistance were mostly parts of R. P. Yadav's PhD dissertation. K. N. Komblas did an outstanding job with his PhD dissertation on field studies of aphid vectors of sugar cane mosaic virus disease in which he showed that most disease spread is caused by the restless feeding of winged aphids of many species, probing the plant leaf for a taste of sap before moving on to something more to their liking. Prior to this, it was believed that only a few aphid species living and reproducing on sugar cane plants were the main vectors. We provoked Dr. Chilton with this since plant diseases were supposed to be his specialty. Yung Song Pan developed improved diets for laboratory rearing of SCBs. The presence of a diapause or a hibernation-like stage in SCB larval development was documented and described by K. P. Katiyar in his PhD dissertation. R. A. Agarwal, in his PhD dissertation research, studied variation in taxonomic characters that define the SCB species, *Diatraea saccharalis*, and distinguish it from its close relatives. Raphael Perez proved that the female SCB moth produced a chemical sex attractant that brought males to her in abundance between the hours of 1:00 AM and 4:00 AM. This study resulted in a grant to my project of $29,277 from the American Sugar Cane League through the LSU Foundation to continue support of SCB sex attractant research in 1965-66.

I developed the reputation for doing a good job in supervising graduate student research programs and spent a lot of time with my students discussing what they were doing, why they were doing it, whether it might be better done another way, what they had concluded from this so far, etc. When it came to writing theses or dissertations, they usually had to rewrite everything several times before I permitted it go to their other committee members. This was partly because most of my students were foreigners for whom English was not their first language. I also insisted that every written description of a completed research problem must end with numbered statements of conclusions that were adequately supported by data and observations presented. There is a tendency among some scientists and many budding scientists to end research reports with a section on "discussion," leaving the reader to wonder what in the world was concluded. I never permitted that, a fact that bothered a few people. However, for whatever reason, increasing numbers of graduate students were assigned to my project until at one time, just before my resignation to go where I could start a consulting business, I was simultaneously supervising as many as six graduate students. This had become almost a full-time job in itself, although it was one that I enjoyed and in which I experienced a sense of genuine accomplishment.

My relationship with these students was much more professional than personal, not that I preferred it that way; that is just the way that it was. We could enjoy conversing for hours in my office about research matters, but in my home at the dinner table or afterward in the sitting room, conversation was usually more strained. Nevertheless, I contributed to their lives in some useful ways, and they all went on to successful careers. Kamta Katiyar became a professor of entomology at a university in Caracas in Venezuela. Yung Song Pan returned with his MS degree to the agricultural experiment station in Taiwan for several years before returning again to LSU for a PhD. R. P. Yadav became a professor of zoology and entomology at Southern University in Baton Rouge. Melvin Kyle completed his MS degree after my departure from LSU and had a long sales career with Velsicol Corporation in the southeastern United States. R. A. Agarwal returned to an experiment station job in India. Howard Anderson had a long and successful career as an agricultural chemical industry technical representative in Texas and other southwestern states. Raphael Perez returned to the agricultural experiment station in Puerto Rico as a research entomologist. Abdel Latif Isa returned to Egypt as head of the entomology department in the Egyptian Ministry of Agriculture and later became first undersecretary in the ministry there. Constantinos N. Komblas accepted employment with Dow Chemical Company, which transferred him to his home in Athens, Greece, where he later assumed technical and research responsibilities for all of Europe, Africa, and the Middle East.

Dr. Philip Callahan taught a graduate course in applied entomology at LSU until 1962, when he resigned to go with the USDA. He had always taught this course more or less as a survey of common pests and their biology and control. Since this was also the approach of an undergraduate course being taught by our department, I elected to make it more a course on principles of applied entomology than a survey of common pests. I wanted to emphasize data collection and the use of numbers in describing pest populations, determining economic population thresholds, etc. I rewrote the course, first teaching it in 1962 and thereafter until I resigned from LSU.

Restless for Change. The workaholic habits, which I had developed during university undergraduate and graduate studies, became only more pronounced during our eight years at LSU. In my mind, I always blamed this on the ever-present need for more income. I was hired with a brand new PhD degree as an assistant entomologist in the LSU agricultural experiment station at an initial salary of $5,200 annually. The following year, I was raised to $6,700 annually. Who could complain about a 29 percent raise? However, there was still a shortage of cash to satisfy our basic family needs, to say nothing of desires. Also, there was little to no time for family to be together and play together. Janice was already working again at the telephone switchboard in the Baton Rouge General Hospital, where she continued for the eight years that we lived in Baton Rouge. I received top raises annually and was promoted through the ranks from assistant entomologist to associate professor and, in 1964, to full professor of entomology at an annual salary of $11,200. There was concern among some about my being promoted to full professor in such a short time. Beside my duties at LSU, I was now a devout and active Mormon with regular church teaching and administrative responsibilities. The combination of these circumstances all combined to cause disappointment and unhappiness in our home.

Jan was one year old when we arrived in Baton Rouge where she went through kindergarten, first, second, and third grades. Janice saw to it that she had three years of dancing classes. During summer vacation periods, Janice worked the 3-11 PM shift at the hospital so she could be home with Jan to go swimming, shopping, etc. in the mornings and early afternoons. The rest of the year, she worked from 7:00 AM to 3:00 PM. On some summer days, Janice would drop Jan off at my office on her way to work. Jan then would entertain herself with chalk at the blackboard or by wandering about the laboratory until time to go home and eat supper, which Janice usually had prepared and left in the oven for us.

By the time of my promotion to full professor in the spring of 1964, I had learned that there were opportunities to earn extra money during the summer by consulting on weekends with sugar cane plantation managers. When I approached Dale Newsom about this, he was adamant in insisting that neither he nor any of his staff should be allowed to do off-campus consulting. He felt that such activities would detract from overall research efforts and the fulfillment of responsibilities to the university. He was absolutely unyielding in his thinking on the subject.

The year 1964 was a very busy one. There were two or three farmer meetings weekly during January and February, a thirty-minute radio interview on SCB control February 17, chairing the committee for Komblas's final oral exam for the PhD degree February 28, an address to the Louisiana Aerial Applicators Meeting at the Jack Tar Capitol House March 2, the annual report to the Sugar Cane Contact Committee March 19, an address to the Sugar Producers Club April 30, and I recorded a radio tape June 29 to be played on farm news programs across south Louisiana. Precious time was taken from routine office, laboratory, and fieldwork, collecting and processing research data, to speak at several parish field day programs during the summer. At this time, I was also Sunday school superintendent for the New

Orleans stake of the LDS Church that required holding regular monthly meetings of the stake superintendency and also attending meetings with Sunday school workers in the various six or eight wards of the stake. In the course of these duties, it was not unusual to be asked to speak at their sacrament meeting services and ward conferences as well, which I did on occasions in Covington, Hammond, and Baton Rouge. There were numerous other job and church-related meetings that took time and at which my presence was expected.

In spite of these many activities, a few notes in my diary indicate that on March 3, 1964, I fixed supper of bacon and eggs for Jan and me while Janice and neighbor, Ann Smith, went to the movie "Cleopatra." On March 4, Janice, Jan, and I attended a rodeo in the LSU Agricultural Coliseum in which the entomology department was also housed at that time. On March 5, Janice and I went to New Orleans with Jack and Ann Smith to dine and to see and hear Perry Como in person. On July 15, Jan and I went to the Gordon Theatre to see "Hey There, It's Yogi Bear." And on December 4, I baptized her at 6:00 PM in the Baton Rouge Ward Chapel, after which she was confirmed at sacrament meeting December 6, 1964. If I did anything else that year with or for my family, besides earning a tiny paycheck, it doesn't show in my diary. A couple of times during our eight years in Baton Rouge, we drove to Long Beach near Biloxi to spend a summer weekend in a motel across the highway from the beach.

I was doing a lot of praying at that time about what to do, when it occurred to me that a nine-month appointment at one of the state colleges would allow me full-time in summer to pursue consulting opportunities. With this in mind, I wrote in the fall of 1964 to Nicholls State College, eighty-five miles south of Baton Rouge in the little historic French community of Thibodaux, to which they responded with the offer of a nine-month teaching contract for $11,200, the exact amount of my twelve-month contract at LSU. Dale Newsom was flabbergasted to hear that I would even consider leaving LSU at such a time when he believed that I was on the "threshold of much more recognition and reward for so much work so well done." He tried very hard to dissuade me from moving, and I am sure that he was sincere in his arguments, but I had made up my mind in favor of more money immediately as opposed to "pie in the sky bye and bye."

On January 11, 1965, Mr. T. M. Barker, president of Valentine Sugars Inc. at Lockport, assured me of his business if and when I began consulting. In the next several months, I signed up Delta Farms at Larose, Robichaux Estates at Labadieville, Gravois Farms Inc. at Paulina, and John Ory at Convent. These people offered to help me sign up more acreage with their neighbors, but I was afraid to take on more than I could handle. Six to seven thousand acres looked like a full-time job to me at that time, and at one dollar per acre—that was a lot of extra money, or so I thought. I was confident that commercial field scouting to determine the need for and the timing of operations for SCB control had never been done properly and that I, as one of a select few, who now had the knowledge and ability to offer this service to growers, also had the desire to do so. I was determined to show that large plantations could be micromanaged by intense and thorough scouting to avoid treating every acre of cropland with insecticide every time control measures were needed. I was certain that production costs could be saved without loss of yield benefits from SCB control. I was correct on all counts, but as later events would show, I underestimated the resistance that would be encountered.

Janice had radical surgery for gallbladder removal March 16, 1965. In those days, it was a six- to eight-inch incision below the sternum instead of a coin-size opening. At the 9:30 AM time of her surgery, I was addressing the Sugar Cane Contact Committee on the LSU campus as per an earlier-published schedule. It seemed to her that my timing was always the worst possible, and it did seem that way to me as well. She developed some infection but was released from the hospital six days later.

By 1963, Dr. Sess Hensley had tired of battling his boss, Mr. Mathes, at his USDA job in Houma and had moved to Baton Rouge and LSU to accept responsibility for the soybean project. Because he had been involved with sugar cane in the earlier endrin battles, he had a continuing interest in the sugar cane program. And so after I left LSU, Sess was transferred from soybeans to the sugar cane insect research project. On March 26, 1965, I met with Drs. Newsom and Hensley and my new graduate student, Abner Hammond, who since has served for many years as professor of insect physiology at LSU. The purpose of the meeting was to discuss future continuation of research on the chemical sex attractant in the SCB. Dr. Hensley would be taking over my project, and Abner would be doing his thesis research on the subject as part of Hensley's project.

On April 12, 1965, I met with Clyde Coreil of the local TV station to produce a short film highlighting sex attractant research at LSU. On April 30, I took and passed the required state exam for consulting entomologists. We attended the LSU Entomology Department picnic May 1. I met with company representatives regarding registration of the new guthion insecticide May 4. On my last official duty day at LSU, May 10, 1965, I attended meetings of the LSU agricultural faculty and of the entomology faculty. The following day, I drove to Thibodaux for house hunting. Janice worked her last day at the hospital May 21. We both were sacrament meeting speakers May 23 in the Baton Rouge Ward. On May 28, we drove to Decatur for a brief visit and to leave Jan with Mother and Daddy for four days. We drove to Dolphin Island the next day to spend a night at the beach. This was the first night that Janice ever had been away from our daughter since her birth January 20, 1956—more than nine years earlier.

10

TEACHING AND CONSULTING IN BAYOU LAND (1965-73)

There were only four buildings on the Nicholls State College campus in 1965 with an enrollment of less than two thousand students who mostly commuted from Lafourche and adjacent parishes. Thibodaux was an old historic town of about 25,000 population. Upon seeing it all for the first time the previous December, Janice's initial response had not been a positive one. However, I promised to build her dream house in Thibodaux if she would support me in this move. We had built two homes in Baton Rouge, the second larger than the first, but neither really large enough for more children. Seeing my determination and with the promise of a larger home in the process, she became resigned to the move. An so we departed Baton Rouge in the spring of 1965 and moved to Thibodaux into a white frame house at 208 Cherokee Street, one block from the college campus.

We bought our first new car, a green four-door Chevrolet Impala sedan, and a new International Scout pickup truck with four-wheel drive in which I would scout about six thousand acres of sugar cane for a dollar an acre that first summer. This would boost immediately my annual income by more than 50 percent. With such incentive, I believed that no problem could be insurmountable. Janice would be able to stay at home now as a full-time mom with no more double duty between the hospital and home, which she had looked forward to. It was only three blocks to Jan's school, where her mother or I delivered her or picked her up on bad weather days and to which she walked to and fro with new friends on other days. I was only one block from campus and two blocks from my office door there.

We attended the Morgan City branch of the church about thirty miles to the west that was at that time the closest LDS congregation. The branch president was a research scientist from the agricultural experiment station in Jeanerette, thirty miles to the northwest. He traveled weekly with his wife and children to preside and conduct services in this small branch of several families that met in a dilapidated rent house. On our first Sunday there, we were careful to step over and around holes and loose planks in the floor of the front porch to avoid falling through. Janice and I traveled about the same distance with our daughter Jan and a friend and faculty associate (librarian) at Nicholls, Agnes Clark. Within a week or so, I was called and ordained as first counselor to the branch president. The half-hour drive through the swamp to Morgan City and church became a familiar weekly event during which we became close friends of Agnes. I would later learn of her nephew, Dr. Jerry L. Ainsworth, who traveled extensively in southern Mexico, Honduras, and Guatemala excavating and studying ancient Mayan ruins and artifacts and who later authored a book entitled "The Lives and Travels of Mormon and Moroni," published in 2000, which I now have read with interest.

I had a whole summer to work with my new farmer clients before college classes started in September. Like those at many small colleges, the administration at Nicholls State dreamed of growing rapidly into a regional university of prominence. They were initially proud of attracting a young full professor of international prominence at LSU to their small campus. In spite of a busy summer schedule with clients, I met at noon July 1, 1965, with Dr. Thomas Stanley, dean of the division of science and technology, and Dr. Donald Ayo, head of the department of agriculture. We reviewed their plans, to which we had agreed earlier, that I would become a member of the agriculture faculty. However, due to recent administrative changes, Dr. Stanley would soon leave for a new job in Texas, and Dr. Ayo would assume responsibilities as dean in addition to those of department head for a time. Also, due to the present temporary allocation of funds with a shortage for agriculture, I would be assigned to the biology department but with teaching responsibilities both in biology and agriculture.

Jan was now nine years old beginning fourth grade and was taking private piano lessons with Mrs. Janice Gee, a popular piano teacher in Thibodaux. My mother gave us her old piano for Jan to practice on, which she happily did for a while. However, her interest waned after a time, as that of many children does, and she had to be encouraged considerably to practice. Nevertheless, her piano lessons continued for three long years until we all decided that life would be better without them. Jan had been upset about having to move from Baton Rouge to a new place where she would have to make new friends. However, she did what she had to do and made several new friends in the neighborhood with whom she often played for the first several years. One was a daughter of Senator Harvey Peltier Jr., one block away, and another was the daughter of our family doctor at that time, Dr. Tom Kleinpeter, two blocks away.

Dr. Ayo was already present in Mr. Barker's office at Valentine Sugars when I showed up for a business meeting with the Valentine boss. Since Ayo was a native of the area, I assumed they were just buddies shooting the bull. I don't recall that Mr. Barker and I discussed much business. What I do recall is Dr. Ayo's comment that Dr. James Peltier, a Thibodaux oral surgeon, was having trouble with trunk borers in shade trees at his home and that Dr. Galliano would appreciate my stopping by to help him with the problem. On the way home, I took a look at Dr. Peltier's trunk borers and called to tell him what to do about it. He wanted me to come out and look at the problem with him, which I did. I told him that a little liquid fumigant squirted in the hole before plugging it would kill the borer and that spraying the trunk twice a year in the spring and fall with an insecticide might reduce the incidence of future problems. He inquired if I had any of those materials, and I did not. A few days later, Dr. Galliano's office inquired to know if I had taken care of Dr. Peltier's problem. I responded "yes" that I had told him what to do and "no" that I didn't do it for him as my time was limited and my business did not offer that type of service.

I first reported to Nicholls State College for faculty meetings Tuesday, September 7, 1965. It soon began to appear that neither the biology department head, Mr. Max Hardberger, nor his boss, Dr. Phillip Uzee, the dean of arts and sciences, had been consulted for their opinions about my hiring. I concluded that the college president, Dr. Vernon Galliano, simply had told them that I was being parked in biology temporarily until funds became available to move me into agriculture. I believe there was resentment toward me for being the highest paid member of the biology faculty at that time, in spite of the fact that I had never taught any of the biology courses and certainly had never taught undergraduate college students before. Therefore, Mr. Hardberger and Dean Uzee decided to put me to the test. I was assigned to teach entomology lecture and lab and an elementary course in experimental statistics called "Field Plot Techniques" with lecture and lab for the agriculture department, in addition to freshman zoology lecture and labs and freshman botany labs for the biology department. The only one of those subjects I had ever taught before was entomology. I had expected to be challenged, but not quite so harshly. My assigned teaching load this first semester amounted to eighteen contact hours weekly, not counting the additional time for preparation and grading.

I worked at student registration Wednesday through Friday, September 8-10. As we finished registering in midafternoon on Friday, radio reports were telling of the approaching Hurricane Betsy in the Gulf of Mexico. The wind was picking up noticeably by 4:00 PM, so I headed to the city dump where I had earlier seen some scrap lumber that I took home and used to board up our windows. According to the radio, this was a strong hurricane, the eye of which would probably pass east of us through Mississippi. However, the rain bands began passing over before dark, and the winds continued to increase. By 10:00 PM, wind velocities were exceeding one hundred miles per hour, and we were holding folded bath towels around the living room window air conditioner in a futile attempt to stop water from entering the house. The transformer on a pole by the street exploded in a shower of sparks, and I glimpsed between the window boards a large airborne tree limb sailing down the street about eight feet above the ground. Shortly after midnight, the winds ceased abruptly. We had not had radio or TV contact for a while, and all phones were dead, but we realized now that the earlier storm reports had been misleading and that we were actually in the eye of the storm. I went out the front door to quickly survey damage, knowing that the winds would shortly start again from the opposite direction. The yard and street were under about twelve inches of water, and broken parts of trees were everywhere. The front porch roof was sagging, and the bases of the four supporting columns were resting on lawn grass just beyond the edge of the concrete porch floor. After about a half hour, the winds resumed and continued throughout the night gradually abating to less than hurricane force by midmorning Saturday. We later learned that maximum winds of one hundred fifty miles per hour had passed through Thibodaux. Telephone service at our home was lacking for more than two weeks. Nevertheless, classes at Nicholls started on time Monday morning.

When I discussed the matter of my ridiculously heavy teaching load with Dr. Ayo, the most comfort he could offer was to "just relax and go talk with those kids about any thing that interests you." The idea that three or four hours credit in any course should represent a certain standard amount of basic knowledge did not seem important to him. His theory seemed to be that anyone could teach any course by simply reading a little from the text and then launching out on any discourse that comes to mind as a result thereof. I was beginning to understand at least one reason why degrees from some universities may be more meaningful than those from others. As disappointing as this advice was, I could not and would not permit myself to follow it. And so I spent that first semester, very much as I had my last few years of graduate school, working until 2:00 AM or later and then catching a catnap before going to meet classes.

I attended the men's faculty club party one evening and seemed to be the only one there not drinking alcohol. I was on the wagon for religious reasons and for life, or so I thought. Lots of booze was flowing; and blackjack, poker, and bouré were the main focus of the evening. There were six or eight players at President Galliano's round table. I hung around like a wallflower for a while before leaving and feeling much relieved to get away as I drove home.

One year after leaving LSU to begin consulting, I was asked to participate in a panel discussion of "professional entomology in Louisiana" as part of the program for the third annual meeting of the Louisiana Entomological Society. The panel was composed of an aerial pesticide applicator, me as a consultant, a research entomologist, and an extension entomologist. By this time, I had worked one summer for a number of farmer clients and had begun to butt heads with the aerial pesticide applicators and occasionally with extension personnel. While we all served farmers, consultants were very much focused on reducing production costs without simultaneously giving up any pest control. Many aerial applicators were focused on giving the farmers enough pest control to keep them happy while flying over as many acres of cropland as possible. Extension Service personnel, including county agents, were a little unhappy with a new and growing group of agricultural consultants who they felt were competing with them to provide services on the farm. Truth was that private consultants were now offering far more service(s) on more acres of cropland than could possibly be offered by a few government employees. In my address, I served notice that we were here to stay and also to help the sciences of agriculture better serve mankind. I talked about the various services that consultants might offer, their responsibilities to their clients and to the public at large, the problems they encounter within their own group and within other groups represented on the panel, and how our working relationships might be improved.

To walk in a Louisiana sugar cane field during the summer months is more physically challenging for a number of reasons than walking in cotton, soybeans, or any other crop, except perhaps a flooded rice field. There is no level ground between two cane rows since the middle ground is a water furrow from which soil is thrown up against the bases of the cane stalks. Long before the stalks reach their final height of ten to twelve feet or more, the leaves from adjacent plant rows mesh together to form a barrier that requires at least one hand to protect the face while walking or fighting one's way through the foliage. Frequent rain requires that scouting often be done in mud or in several inches of water on occasion. Although not encountered daily, a field scout must keep an eye out for an occasional encounter with a poisonous snake and, less often, with coyotes, bobcats, and even alligators.

One summer afternoon, I was scouting cane on a ridge behind the Valentine Sugar Factory near Lockport, Louisiana, when I became aware of something unusual from the corner of one eye. Looking more directly at it, I realized that I was standing motionless with a foot on either side of a very large snake that had been crossing the middle between two rows of cane. Neither the reptile's head nor tail was visible, although the cane rows were almost six feet apart. Aided by a sudden surge of adrenalin, I jumped at least six feet away from the critter as it quickly coiled and faced me. After calling my helper to bring my shovel from the truck and to take my place, he kept the snake's attention while I circled around the other side and whacked its head with the shovel. With the reptiles limp body draped over the shovel, I carried it out of the field and threw it in the back of my International Scout pickup to take it home and show to the family. I thought that everyone living in this part of the world should know what a canebrake rattlesnake looked like. As we drove home, I glanced through the rear view window occasionally to be sure that all was well. The third time I looked, the rattler had revived and was crawling about in the truck bed. Before it was dispatched a final time at home, my family had seen a real live and angry canebrake rattler almost six feet long with its dark tail and rattlers, body as big around as my ankle, and fangs about two inches long.

During thirty-five years of field scouting, I or my employees killed only three snakes like this, although they were often found dead on field roads where they had been run over by trucks. Although a dangerous snake, due to its large size and venom, it must be less aggressive than the more commonly encountered cottonmouth moccasin. Otherwise, I might not have been so fortunate for so many years. Following the final spring field cultivation, snakes move increasingly into cane fields in search of rabbits, rats, mice, quail, or other prey. I'm sure I've slept better by not knowing how many times I've been within striking range of a canebrake rattler. Cottonmouth moccasins were more commonly seen, particularly after heavy rains, sometimes even in the cane several feet above the ground, apparently to get out of the flooded middles. Why, I don't know. They are usually found in water or basking near it where they will eat any vertebrate prey small enough to be swallowed, although fish are the main items of their diet. The cottonmouth was the most frequently seen poisonous snake and apparently the most ill-tempered. On one occasion, I saw a cottonmouth still alive in a drainage ditch where it had tried unsuccessfully to swallow a catfish, two fins of which were protruding through the snake's neck.

August 26-28, 1966, Janice and I with daughter Jan, Janice's aunt Chloe, and my mother traveled in our 1958 Buick to Salt Lake City where Janice and I were married for time and eternity in the Salt Lake Temple and where we received our endowments August 30 of that year. During two days there, we also toured the Tabernacle, visited the church's genealogical library, the Great Salt Lake, and the This is the Place Monument. On the return trip, August 31-September 4, we traveled through Arizona, New Mexico, and Texas visiting Zion Canyon, Bryce Canyon, the Grand Canyon, the Petrified Forest, Carlsbad Caverns, the Painted Desert, the LBJ Ranch, and other points of interest. Aunt Chloe and Mother maintained a lively conversation during most of the trip.

In late winter of 1966, Dr. Ayo insisted that I address the Thibodaux Women's Garden Club on general gardening with emphasis on garden pests and their control. This was not a subject that I had ever been much concerned about before, but I got together some slides and borrowed some others for a presentation that I made without feeling very comfortable or happy about. I began to get invitations to speak at Kiwanis and Rotary luncheons and other functions about the subject of SCB control and previous research on that subject, which I did know a great deal about, but always made excuses to avoid such assignments. I would have done battle if necessary with Goliath on that subject, but to voluntarily talk on any subject to a group of community leaders in a social setting at that time would have been more trauma than I was willing to volunteer for. In these and in many other instances, I failed to take advantage of opportunities to bond personally with community leaders, many of whom were involved in sugar production. I thought that I already had all the business I could handle anyhow. However, I did meet with the Terrebonne Parish Police Jury, February 24, 1971, to answer questions they had about controlling fire ants in their parish, as one of my customers was a member of the police jury at that time. I convinced them to ban the broadcast application of Mirex insecticide for fire ant control to the considerable displeasure of the state entomologist, Richard Carlton, who spoke there in favor of the program.

Dr. Ayo later approached me about teaching a course in soil science, which I was qualified to do by training but unwilling to do as an add-on in addition to everything else I was then teaching. Also, a couple of years had now passed, and I was still in the biology department where my salary had begun to lag while the rest of the crowd was catching up. I informed him that I would be unwilling to accept any additional course responsibilities unless or until I was released from some present ones and that I was unwilling to teach in more than two subject matter areas in any one semester. This conversation may have helped seal my fate at Nicholls as Dr. Ayo later succeeded Dr. Galliano as president of the school.

On November 15, 1966, I received an unsolicited phone call from Dr. William Luckmann, head of the Entomology Department, University of Illinois at Urbana, offering me a job on the recommendation of Dr. Dale Newsom and describing a new faculty position with responsibilities for developing cooperative research and extension programs in integrated pest management (IPM) with third world countries. He wanted me to visit and discuss my possible interest in the job with him and his staff. However, the best annual salary he could offer was $17,000, not quite as much as I was then making from combined teaching and consulting activities. I was flattered by his interest but declined the invitation. A few weeks later, a letter came from North Carolina State University in Raleigh inviting me to apply for a faculty opening in their cotton entomology position. I did not pursue this for the same reason I had not pursued the Urbana job.

In early 1969, the church authorities organized a branch of the church in Thibodaux. I was called as first counselor to Ronald Ury, an engineer from Salt Lake City hired by the Thompson Machinery Company of Thibodaux to supervise their operation of manufacturing sugar cane farm machinery and equipment. We first met in Ury's home and, after that, in a small rent house. Later, I arranged for the church to rent the home of the newly appointed president of Nicholls State University as a meeting place for the branch. Dr. Galliano had recently moved to the president's home on the Nicholls campus and was happy to rent his home to a trusted faculty member. A year later, Ronald Ury's job with Thompson Machinery Company ended, and he with his family returned to Utah. I was then called as branch president to assume responsibility for a growing church branch.

My departure from LSU in 1965 to begin consulting and the organization of a state agricultural consultants' association the following year marked the beginning of a period of competition between the LSU agricultural experiment station and extension service (including county agents) on the one hand and agricultural consultants on the other. Prior to this time, most field crops had been treated with insecticides by aerial applicators on some sort of automatic schedule after the initial need for treatment was established. And that need had been most commonly identified for large areas over many fields covering many hundreds or thousands of acres. I was determined to show in sugar cane that this could be done field by field through intensive weekly scouting to dramatically reduce crop production costs and adverse effects of needless pesticide use. No county agent or other government employee had the time or inclination to provide this kind of service. I was so successful in doing this that several aerial applicators became my enemies overnight, doing all they could to discourage farmers from using my services. Sess Hensley told me a few years later in 1976 that the Extension Service had kept records for a while on the amounts of insecticide treatment done on sugar plantations serviced by different consultants and that my clients were using dramatically less insecticide than those of any other consultants. He said that he had urged them unsuccessfully to publish the information as a public service but that they would not do it.

On Sundays, my church leadership meeting started at 7:00 AM and was followed by priesthood, sacrament, and Sunday school until noon. After that, it was grab a sandwich, if luck permitted, and then interview or visit with individual members or groups the rest of the afternoon. Also, I attended monthly stake leadership meetings in New Orleans on first Saturdays that often lasted most of the day and into the evening. Too much to do with too little time? I began to wonder.

During the next two years, our congregation grew from five or six families to more than a hundred members. Many of our members were new converts. Others were transients, and a few were inactive and becoming active again. Most traveled from fifteen to as much as fifty miles one way to get to church. My two counselors in the branch presidency were very helpful, but there was more work to be done than people to do it. I was trying to do more than possible and didn't know how to stop this ever-escalating work schedule.

On one occasion, the Mormon missionaries were arrested for tracting from door to door in violation of the Green River Ordinance prohibiting door-to-door selling. They called me from the police station. I called and met with Mayor Harang, who was also a large landowner and sugar cane planter with sincere respect for my contributions to the region's agriculture. He explained that many of the Catholic citizens, which included himself and the great majority of other people in Thibodaux, were upset by these young men traveling on bicycles and knocking on doors to "sell" their religion in this conservative small longtime Catholic community. After discussing the matter, he agreed to get the police chief off their backs if they would not revisit any addresses after being told that the occupants were not interested—a very reasonable request to which I happily agreed. Besides assisting the young missionaries with various other problems, they frequently were visitors in our home, especially at meal times. We enjoyed their attention and they ours for quite a while.

Facing large numbers of college freshmen on a daily basis was a new experience for me. I had never considered that motivating students should be a necessary part of teaching. Graduate students had always seemed anxious to receive anything I had to offer. But now with fifty to more than a hundred freshmen students in a classroom or auditorium at 7:00 AM, or any other hour of the day—having to listen to me characterize the different animal phyla and explain why one is more primitive than another, and how it probably got to be what it is through natural selection acting over long time periods upon large populations, whose variations could partially be explained by Mendelian principles of genetics, the basics of which they must try to understand—there just weren't many of them in the mood for all that "hogwash."

I began trying to at least start my lectures with something that might be funny or interesting to the majority, a short joke or some special news item. On a couple of occasions in sheer desperation, I showed them a dance step I had learned at Arthur Murray's Studio. I remember that did get their attention one day. For my first couple of semesters, the back rows were always filled with athletes. However, that changed by the next year when it had become general knowledge that I would not respond to requests for grade changes, regardless of their source. Occasionally, a smart guy, causing continual disturbance, would be ordered out and told not to come back. On one such occasion, the boy refused to leave, saying that he had paid his fees and had every right to stay. I told him he had lost that right by disturbing the class a second time after being cautioned. I walked to his desk and picked up several of his books before walking to the door where I threw them into the hall and ordered him to follow. Fortunately, I suppose, he did. However, Dr. Green, department head at that time, cautioned me to call Campus Security if there were such incidents in the future. No other student ever refused to leave when asked to do so.

Over the years, I earned a reputation with freshman students for being uninteresting but thorough and hard but fair. Occasionally, a student would express appreciation for what he or she had learned in my class, which helped make it seem worthwhile. One artistic middle-aged mother of grown children, upon completing my freshman zoology course with a top grade, presented me with a large framed woodcut depiction of the phylogenetic tree of animal life and thanked me for helping her to understand something that had never made sense to her before.

My students in entomology were upperclassmen, better motivated, and more interested in the subject than freshman biology students. I taught one entomology class annually. A significant percentage of those students went on to graduate studies elsewhere and several did so in entomology. Two out of ten ag students failed my course in field plot techniques (statistical methods) my first year at Nicholls, after which I refused to teach the course again. One was the son of a large customer of mine. He was a likeable young man and extremely popular with everyone, but books simply were not his favorite thing. He came to my house to plead his case, but his final grade had not even been close. I was genuinely sorry for him and for the fact that his father was a very important customer. But, as a matter of principle, I could not and did not pass him. His father and I never discussed the matter, and I continued to work for him and to be on good terms with the young man until his untimely death eight or ten years later.

In about 1966 or 1967, the college scheduled a Leap Week during March. Nineteen professors, local ministers, priests, and students spoke from their viewpoints on the concept of God in today's world. Is God dead? Is organized religion relevant? How does the student reconcile his childhood religion with his adult profession and responsibilities? On March 4 at 10:00 AM, I spoke to a sizeable audience of students and noncollege citizens on the subject of "religion and science," with special attention on the subject of "evolution." Given the fact that individual evolutionary events are often random, similar to the appearance of heads or tails in the tossing of a coin, I maintained that the end results of evolution, set in motion by an all-wise creator, are at least as certain as the fifty-fifty probability of heads and tails from many fair tosses of a coin.

The first Earth Day celebrations also were held on campuses throughout the country during a year in the late sixties. Religion, morality, environmental concerns, pesticides, authority, etc.—all were being scrutinized, questioned, and heatedly debated. There were several new faculty members in the biology department now, one of whom, Dr. Curt Rose, was a bright and talented somewhat younger research scientist and also ultraliberal on almost every issue. Curt and I were natural opponents on most subjects, including religion and pesticides. He was an outspoken atheist and protector of the environment at all costs. I, as a Christian, agriculturalist, and environmentalist, believed that proper use of pesticides was absolutely essential to affordable crop and food production and even to proper maintenance of the environment. We both enjoyed friendly discussions of our differences for a time until our discussions became so heated that he became highly insulting on an occasion prompting me to order him out of my office. We were invited to be featured speakers at the Earth Day celebration in a debate involving religion, environment, and pesticides. No official winner was declared, but each left the debate confident that he had won the day.

In 1966, I was invited by the International Society of Sugar Cane Technologists to write, for a book entitled *Pests of Sugar Cane*, a chapter on the subject of "Insecticidal Control of Moth Borers of Sugar Cane." Reviewing literature on any subject via the Internet was not yet a reality, and Nicholls had no adequate research library. However, I accepted the invitation, without any reduction of teaching load or other assistance from NSU, and did legwork to and from LSU eighty miles away. The chapter was written, and the book was published in 1969 by the Elsevier Publishing Company. In 1968, I was president-elect of the Louisiana Entomological Society and president the following year. NSU president, Dr. Galliano, noted this in a congratulatory letter to me in 1969 in which he stated that he was "well-aware of the work you have been doing, and I appreciate your efforts."

An editorial appeared in the *LDS Church News* about this time condemning teachers of evolution as "sons of perdition." I responded with a letter to the editor reminding him that the Mormon Prophet Joseph Smith had advocated seeking knowledge from all good books and asking if that did not include those from the biological sciences. My letter was sent to the general authorities of the church who responded by way of a letter from Apostle Mark E. Petersen to my stake president, Dr. Melvin Gruwell. Dr. Gruwell showed me the letter at our next meeting by way of informing me of the LDS Church's position on the matter. I thanked him and told him that my thinking was not altered by the letter, and we both let the matter drop.

In 1970, the editorial committee for the *Annual Review of Entomology* invited me to write a chapter on "Insect Pests of Sugar Cane" as a review of world literature for the 1972 edition of that book. This was a high honor for my profession and one that I could not easily turn down. Nevertheless, I was carrying a fifteen-contact-hour teaching load, with no chance of that being reduced, and was still far away from a good research library. Therefore, I asked Sess Hensley, who was still on the LSU faculty at this time and had been an earlier research partner, to join me as junior author in this effort. He gladly agreed, and between the two of us, we authored a twenty-seven-page review, including 171 literature citations, describing much of the then-current knowledge of the biology and control of sugar cane insects of the world. Two years later, I was included in the 1971 edition of *Outstanding Educators of America* "in recognition of contributions to the advancement of higher education and service to community." And in 1973, Marquis Who's Who Inc. included me in their fourteenth edition of *Who's Who in the South and Southwest*.

My consulting business, which until now had sold information and recommendations only, based upon weekly field scouting and survey information, was in its sixth year in 1971. We were attempting to expand into urban pest control as Long Pest Control Inc., providing termite, household and ornamental pest control services to permit year-around employment for some workers. At the same time, I was a full-time faculty member at the local university. I was wearing at least three hats outside the home;

and we had three children at ages fourteen, three, and nearly one year old, who now had little chance to get as much of their father's attention as they deserved.

Birth, Death, and Marriage. Daddy had been in the hospital for some time with a chronic urinary infection, and Mother called to say that he appeared to be failing rapidly. I flew home in early November to visit him in the hospital for three days. Several days after my return to Louisiana, I was told that he got dressed and left the hospital, against doctor's orders, to go home and lie in his own bed. I flew home to visit again in late November and tried helplessly to cheer him by recalling some quail hunts we had made together, but to no avail. He told me he had hoped to leave his three sons an inheritance and was sorry that he couldn't. He said that he was obliged to leave what he had to Mother and hoped she would not give it all to charity or church before her time expired. Mother called again December 3, 1965, at 6:00 AM to say that he had died several hours earlier. Janice and I with Jan drove all night to arrive in Decatur at 5:00 AM December 4 and to attend Daddy's funeral Sunday, December 5, before returning to Thibodaux the next day.

We had moved to Thibodaux for more money and a larger home. Janice planned her two-story dream house, and I bought three large lots on the bank of Bayou Lafourche in 1966. In June 1969, we moved into our two-story, four-bedroom, three-and-one-half bath, forty-five-hundred-square-foot dream house on the bank of Bayou Lafourche two miles south of Thibodaux. We had lived in the small, white, two-bedroom, one-bath, frame house at 208 Cherokee Street near Nicholls for two years before selling it to buy three adjacent lots on the bayou side totaling almost one and one-half acres for the price of $7,800. We had rented a small two-bedroom brick house on Camelia Drive for two years while our dream house was under construction. Janice had saved magazine pictures for years and had a good idea of what she wanted. With her help, Mr. Dooley Hebert drew up our plans, which were sent out for bids, and the contract was awarded to Mr. Louis Hebert, a well-known builder in the area at that time, who was reputed to do outstanding brickwork. The house was centered on two of our three lots and built of Chicago old brick with two large porches, front and rear, and with a forty-foot kidney-shaped swimming pool added the following year.

We had decided, before building, that such a home would be wasted without more children to fill it. Jan had always wanted brothers and sisters, and as a young and able Mormon couple, we felt a duty to help populate the earth as well as the church. And so we were ecstatically happy on the arrival of our second child and daughter, Nancy Anne, December 16, 1967, for whom we had waited a long time. Realizing then that we still had plenty of room, we welcomed our first son, Daniel Henry, to the family circle January 13, 1970, and, twelve months later, William Pratt, who was born prematurely January 21, 1971.

One evening, I drove with daughter Jan to a local unlighted airstrip and turned on my headlights for Hall Lyons to land his piper cub. He was an oil patch millionaire and LDS member campaigning as candidate for governor of Louisiana. He had flown from the northern part of the state to speak to our young people at the church that evening. His meeting with our MIA (Mutual Improvement Association) young people was a great success. Hall's talk and the refreshments after were enjoyed by all, as was the opportunity to shake hands and talk personally with a potential governor. After returning him to his airplane, parked on a dark airstrip at Schriever, we hurried home to arrive at nearly 11:00 PM.

Upon arrival, Janice was in serious trouble. Her water had broken at the end of her fifth month of pregnancy. Nearly-one-year-old Daniel and three-year-old Nancy were in bed. I rushed Janice to the Thibodaux General Hospital, where Dr. Kleinpeter, our family physician, soon summoned, at Janice's request, Dr. Hansen, our pediatrician. The two doctors worked together for several hours. At 5:00 AM, January 21, 1971, William Pratt Long was born alive but with poorly developed lungs. To all appearances, he was perfectly formed. However, Dr. Hansen gave him only four or five hours at most to live and strongly suggested that Janice not see him again or handle him. She had seen him briefly following birth before he was taken to the incubator. The doctors believed she would suffer much less emotional trauma if she followed their advice. She was very reluctant to do so and wanted badly to hold her baby, but the doctors and I prevailed, fully trusting in their wisdom.

All vital signs in William Pratt's two-pound body had ceased by 9:00 AM. I gathered him in my arms, took him to Landry's Funeral Home, and requested he be dressed, covered with his blue blanket, and placed in a white wicker casket. Arrangements were made quickly with my mother to have a little grave opened beside Daddy's headstone in the family cemetery plot in Decatur, Alabama. It was decided that I should depart Thibodaux before noon with William Pratt and make the ten-hour drive to Decatur that day so that a private family burial could be done the next morning. My first counselor in the church branch presidency, Glen Smith, insisted on accompanying me out of concern that I might sleep at the wheel. We arrived at 810 Jackson Street in Decatur at about midnight. My parents had moved back to town in the late 1950s. Mother was up waiting with the covers pulled back in her guest bedroom. I placed William Pratt on the bench, which is now in Nancy's house but then was in Mother's living room, and slept on the couch beside him. The next morning, we took William Pratt to the cemetery where we held a private service in which I dedicated the grave and read scripture and Glen prayed. I placed the casket in the grave where we waited to see it closed and then began our long and tiring ten-hour drive back to Thibodaux. I would soon request and be released June 18, 1972, from the calling of branch president of the Thibodaux branch of the church.

Our home on the Bayou was the scene of a beautiful wedding and reception for our daughter, Jan, who married Denis Smith, Glen's son, August 31, 1973, during her senior year. She graduated from Thibodaux High School in January 1974. In my diary for that day, I wrote "Today Janice Faye Long, my oldest daughter," but never finished the sentence. She had been the tallest and prettiest girl in the Tigerette dance squad her junior year, for which she was ineligible as a senior due to her marriage. Except for me, consensus was that, with so few Mormons to choose from, neither Jan nor her sweetheart would ever have a better chance than this to marry happily in the church. Jan had not enjoyed her school days. She had only one really close girlfriend, Fanny Naquin. They both had wanted to join the navy their senior year, which I refused to permit her to do. She presented us with three precious grandchildren—Holly, Denny, and Heather—before this marriage ended. Holly was born January 4, 1975; Denny, November 24, 1975; and Heather, March 6, 1978. We helped raise all three children while Jan worked and attended college and until she remarried in 1981.

11

Egyptian Mission (1973)

My business in the early seventies was funneling cash into the expansion of pest control services at the expense of consulting income, and we needed more cash inflow. Consequently, when an opportunity was offered me by the Food and Agriculture Organization (FAO) of the United Nations to go to Egypt for four months at nearly twice my current college salary, I accepted the invitation, believing it to be a fortunate and timely opportunity. The university gave me a semester's leave without pay. My employee, Charlie Taylor, could handle the business, with additional help if necessary, for a brief period. Janice with two children, ages four and six, was left to "keep the home fires burning."

I departed New Orleans for Cairo, Egypt, via New York and Rome at 9:30 AM, September 15, 1973, and arrived Rome at 8:20 AM the next morning aboard a gigantic, two-story, jumbo jet. For souvenirs of Rome, I purchased in the airport terminal two artistic eight-inch-tall ivory sculptures of a nude man and woman, which have now graced our bedroom for many years, before taking a cab to the Hotel Pensione Sant An Selmo. The pensione was a family operation in which Papa ran the office, Mama the kitchen and dining room, with children and hired help doing the rest.

Everything was worn but sturdy and clean, including hardwood floors, area rugs, furniture, beds, and linen. Guests dined together at the same time and at the same large table in the dining room that opened by French doors onto a patio. A nice breeze blew through the doorway during meals while several cats from the garden walked about under the table rubbing against the legs of the guests. After lunch at the pensione, I walked in the afternoon to see the FAO headquarters building, only a few blocks away, where I would be briefed the next day. Then I hiked around to find some of the ancient ruins including the Arch of Constantine, the Coliseum, the Palatine Hill, and the Roman Forum. Standing inside the Coliseum and contemplating the horrors that had occurred there as public spectacles nearly two thousand years before was an emotional experience that caused a few goose bumps.

The next day was spent with FAO administrators reviewing my contract, getting travel authorizations, checking medical clearances, etc. I had signed a contract the previous June to work as a consultant for the Plant Production and Protection Division (FAO) in the Arab Republic of Egypt for four months from mid-September 1973 until mid-January 1974. My duties were to work in collaboration with the FAO project manager and the counterpart Egyptian staff to review the present status of sugar cane pest control and to advise on a research program for the assessment of the economic importance of sugar cane pests and their control in Egypt. My main collaborator of the counterpart Egyptian staff was Dr. Abdel Latif Isa, who had done his MS and PhD research under my direction at LSU. My salary was $2,000 monthly with a daily subsistence allowance of 5.93 Egyptian pounds.

The following day, I returned to FAO headquarters for further briefing, travel advance and payroll matters, and to pick up my airline ticket with added baggage allowance. I also mailed to

Janice and children my first audio tape, reviewing the trip thus far, before boarding a plane for Cairo at 5:30 PM. We stopped in Athens, Greece, for a half hour before arriving in Cairo, Egypt, September 18, 1973, at 10:45 PM. After getting through customs, which entailed a search of my person as well as examination of every baggage item, I arrived by taxi at the Mena House Hotel near the pyramids three hours later. I was exhausted and felt a little hassled by customs but was dazzled by the luxury and splendor of the Mena House. After crashing in a luxurious bed, I went to sleep wondering what my next address in Cairo might be like.

Upon awaking a few hours later, I called and talked to Janice and the children and then tried unsuccessfully to call anyone with the FAO in Cairo or with the Egyptian Ministry of Agriculture who might know about my arrival. After a while, I decided to just relax and enjoy. Isa finally called shortly after noon and arrived later to help move me and my luggage to the Mayfair Pensione Residential Home in Zamalek, located on an island in the Nile River and separated from downtown Cairo by the July 26 Bridge. We had dinner at his club before visiting Mrs. Isa and their two daughters at home for a couple of hours. We walked a short distance to the home of Dr. Mostafa Hafez, an administrator in the Egyptian Ministry, where we visited for another two hours before I took a taxi back to the Mayfair Pensione.

The Mayfair was a popular and reasonably priced place to stay for short-term residents and international travelers who were always coming and going. I was the only U.S. citizen living there during the nearly four months of my stay. Most U.S. tourists would have preferred something more comfortable and attractive. However, it was a popular place for people on business trips, government employees, scientists, teachers, and the like, particularly from Europe and the Middle East. Air-conditioning was by ceiling fans and open gas heaters in the dining and sitting rooms. A single toilet and bath with shower was located at the end of the hall on each of two floors. The dining room was furnished with eight plain wood tables covered with oilcloth, each with four unpadded wood chairs. The room opened on to an uncovered second-story veranda. Waiters were all male, dressed in white, and spoke little or no English. The kitchen and sitting room adjoined the dining room on the second floor. Most guest bedrooms were on the third floor while the manager, Mr. George Ghanem, occupied the first floor of the building, which had a gray concrete exterior. The interior had not been painted for a long time, and most rooms were lit by a single bulb or fixture in the middle of the ceiling.

The next day was my forty-fifth birthday. Dr. Don Stewart, my FAO project manager, picked me up at the Mayfair and took me to the United Nations Development Program (UNDP) headquarters for routine business before accompanying me to meet Dr. Moursi, FAO deputy regional representative. He was trained in biological control, was highly critical of insecticides generally, and suspicious of those who sold or promoted them. He terminated our visit shortly after I made a statement in defense of insecticides and their proper use in integrated pest management programs. I was then introduced to other scientists with the UNDP, before attending a two-hour session of the Mid-East Crop Protection Conference, at which Dr. Isa introduced me, to my embarrassment, as "one of the best entomologists in the world." I certainly knew better than that, and surely he did also. I concluded that such exaggeration is not uncommon in some places. In spite of the fact that I was not a cotton entomologist, I was invited to participate in a roundtable discussion of cotton insect problems that apparently were related to the overuse of pesticides in automatic treatment programs without sufficient pest population monitoring by adequately trained personnel. I tried to make a few helpful comments that were accepted politely.

The next day, Don with Marion, his wife, and Bonnie, his daughter, picked me up at the Mayfair for a tour of Cairo that lasted until 6:00 PM when I insisted on going back to the Mayfair rather than to their home for sandwiches. I felt that I had burdened them enough for one day. We had visited the palace of the real Mohammed Ali and the nearby mosque and photographed Cairo from the heights near the palace. We had lunched on shish kebab at a restaurant before spending the afternoon touring shops in their favorite area of Cairo. I had become most interested in a ring, which a jeweler friend of theirs, Mihran Yazajian, was making. When completed, it would have a central diamond with numerous smaller ones surrounding it in the form of a pyramid. I also learned that letters and postcards were censored and did not always go to their intended destinations and that picture taking was restricted and made more difficult by the shortage of film.

It became the custom for Don to pick me up each morning on our way to the ministry and to work. During the next few days, I attended additional sessions of the crop protection conference, visited Stewart's home for dinner with some of their friends, met Mr. Abraham Boulos, director of Experiment Stations, and went to his home for refreshments and a visit with his family. Dr. Ahmed Negm, who had earned his PhD under Dr. Hensley at LSU and was now a professor at Asyut University, invited me to present a seminar at his school. I was invited by Dr. Mohamed Moheb Faki, minister of Agriculture and Agrarian Reform, to attend a 10:00 PM, Sunday dinner at which he entertained the delegates to the crop protection conference. Dr. and Mrs. Isa and I went together to the event at the Semiramis Hotel where dinner was served on the roof, overlooking the Nile, while an Egyptian belly dancer performed accompanied by a live orchestra. She truly had moves I have never seen before or since. I concluded that the movements of her lower trunk muscles must closely approach in speed the wing muscles of a housefly.

During the next week, a lot of time was spent at the office discussing a field survey that I planned to make throughout the sugar cane-growing area. It was decided that I should survey cane fields from Alexandria in the north to Aswan in the south with emphasis on assessment of crop infestation and damage by a species of cane borer that had been known to occur in the Nile Delta and middle Egypt for many years and had recently invaded the southern parts of the country. We planned in detail how field samples of cane were to be collected, how many samples would be examined at each location, and how many locations I would visit. Dr. Isa would accompany me for part of the trip, after which his assistant, Bill Awadallah, would take his place to complete the job before returning to Cairo. By September 30, we completed the itinerary, and now were waiting only for my travel permit to arrive.

In the next few days, I got to know Bill Awadallah, a Christian member of the Coptic Church, which represented a small minority of Egyptian Arabs. Bill was always anxious to talk about religion, and I didn't fail to oblige him. I almost always had breakfast at the Mayfair Pensione, sometimes lunch as well, and usually dined there in the evenings unless invited elsewhere. No more than a mile walk from the Mayfair would put me in the middle of downtown shopping, and I often walked for exercise in the afternoon or evening. I recorded voice and other sound tapes for the family's benefit and wrote Janice weekly. However, she didn't receive any of my letters for the first several weeks. One evening, I wandered into a perfumery where I bought a variety of different perfume concentrates, thinking that Janice might enjoy experimenting with them. Another day, I went with the Stewarts to look at the pyramid diamond ring that was under construction.

I went with the Stewarts to dinner at the home of Mr. Aziz, who lived on the top of a building overlooking Cairo. There I was told that the biblical "Land of Goshen" was near the pyramids at

Cairo and that the site where Joseph, Mary, and the baby Jesus stayed for a time in Egypt was near here. Bonnie and some of her friends took me to see several old churches, including the Church of Abu Serge, the oldest church in Egypt, dating from the third century and built over the site where the Virgin and Christ are believed to have lodged during their flight into Egypt. I went with Marion and Bonnie to the Khan el-Khalili Bazaar and made a down payment at Yazejian's jewelry shop on the pyramid diamond ring for Janice.

The morning of Saturday, October 6, was spent at my office in El Dokki, where the ministry buildings were located just west of the Nile River. The major portion of the city was on the east side of the river and the Mayfair was in Zamalek on the island in between. I mailed a letter to Janice and a postcard picture of the Church of Abu Serge to her and to Mother, before returning to the Mayfair for a short nap and then a walk downtown. Upon returning at about 6:15 PM, I was informed that Israel had been fighting Arabs on all fronts since 1:30 PM that day, that bombing was in progress by both sides, and that our UN district officer, Dr. Ozawa, had ordered that all non-Arab employees stay close to their residences to avoid possible mob violence. I spent the evening in my room wishing that I had a shotgun and a box of shells.

After breakfast the next day, I disregarded the order to stay close to home and walked to the UNDP building to look for mail. I would do a lot of that before receiving any mail from home. The newspaper described Egypt's forces crossing the Suez and taking positions in the Sinai with solid support of Egypt and Syria by all Arab countries. Marion called to say that, for obvious reasons, we would not be able to attend the Coptic Church wedding to which we had been invited. Dr. Ozawa visited again to say that starting tomorrow, we should go about our duties as normal, although the Cairo airport would be closed indefinitely to all traffic and mail would not be going through. A nighttime blackout was now in force, and auto headlights were painted blue.

During this week, word arrived on Monday that my permit to travel had been granted and on Tuesday that it had been cancelled. Isa and I began planning a field survey to be done by sending men to collect cane samples from the fields and bringing them back to Cairo for examination. I worried that the samples would not be randomly selected in my absence and therefore not representative of actual conditions.

Feeling terrible for Janice and the children, who I knew were worrying needlessly, I tried Monday without success to cable and to call and was able to send a cable on Tuesday to alleviate her fears. Thursday, I walked a couple of miles to place a pay-in-advance phone call to Janice for 5:00 AM the following day. Getting up at 4:50 AM, I sat by the phone until 7:40 AM when the call finally came through. It was so good to hear her voice, although she said that no paycheck had yet arrived from UNDP. I wondered how in the world I had called home so easily that first day from the Mena House Hotel.

On Tuesday, afternoon air raid sirens were turned on, three explosions were heard in our neighborhood, and I saw two puffs of smoke high in the sky over the Mayfair, but no planes. On Wednesday, I registered with the U.S. Embassy and declined to be included as a passenger in a truck convoy to help foreigners escape Egypt. Life where I was went on much as usual, although soldiers were dying less than a hundred miles away. Those interested in getting away were primarily tourists while working foreigners mostly were continuing their jobs without obvious concern. Also, I reasoned that returning home at this time would not be financially advantageous and that it would be better to seek to accomplish the FAO mission and also to obtain the diamond ring.

On Wednesday, October 10, I wrote in my diary: "Suspect that none of my mail has gone through—my heart aches for my wife and children—no air raid today—if war goes badly for Arabs, I'm not sure how things will work out here for Americans—more leisure than I've had in twenty years—was nice at first, but beginning to get monotonous—reading is always profitable until eyes get tired—trying to learn how to meditate better—never had much time for it before—have read Exodus, Leviticus, and starting with Numbers now—no wonder there are so few references to these books!"

The Polish embassy evacuated its women and children by bus to the seaport at Alexandria. I watched them load the buses in the next block while waiting at the Mayfair for my ride to work. The morning was spent at the office where an air raid sounded at 1:25 PM following an explosion of a bomb perhaps a half-mile away. Everyone went to the first floor of the building for about ten minutes until an "all clear" was sounded. A notice in the English newspaper advertised an LDS sacrament service to be held the next day at 6:30 PM in an apartment in Zamalek not far from the Mayfair. I checked the address on this day and again a few weeks later, but found no one there knowing anything about it.

At the office, now our conversation turned to the current crisis with Israel. As always, the war was about expanding Israel's borders, this time to include the Sinai Peninsula. I had been furnished earlier with pamphlets and publications documenting by words and photographs the atrocities of Jews against Arabs, beginning with the initial establishment of Israel's boundaries in 1947. The more I read and heard, the more sympathetic I became for the Arab people and the more convinced I was that the U.S. position of one-sided support for Israel must be due, at least in part, to ignorance of the facts by many Americans and to a misguided allegiance of many Christians to the support of Israel at all costs.

In mid-October, I was informed one evening at the Mayfair that the U.S. Embassy had called to enquire if I would like to leave the country by ship from Alexandria. Apparently, someone had thought that I was a tourist wanting to get out. My walk took me that evening to a shop in Zamalek where I bought a lamp with a many-colored glass shade to give to Dr. and Mrs. Isa in celebration of their twenty-fifth anniversary. The next morning, Isa and I visited Dr. Rao, the plant breeder, at his office in Giza, to see if we could use some of his field variety plots and sugar cane seedlings in future research. He was happy to cooperate. In the afternoon, we listened to a live English translation of President Anwar Sadat's speech to his people. Even in translation, it was an impressive and moving oration.

One day, there were air raid alarms at 7:00 and again at 8:00 AM, but no bombs were heard. Bill Awadallah and I spent the morning recording borer infestations in plots of different sugar cane varieties. At lunch in the Mayfair, I met for the first time Dr. Tolstoy, a Russian physicist. His broad knowledge of science in general was impressive, and his knowledge of languages was amazing. He spoke fluent English, French, German, and Russian, all of which I would later have occasion to observe, as he mixed and mingled with other international guests at the Mayfair. At the office, while discussing research with Isa and Bill, I reveled in the plans that were taking form. I was beginning to sense some impending achievement despite being confined to the city of Cairo during a war. I was being carried away with plans for research that sometimes made sleep difficult. At the same time, I was very homesick. If the mail would just start coming through, I thought it would be a lot easier.

A Saturday morning was spent at the office planning, writing, and discussing research with Bill Awadallah. That evening, I attended a birthday party at Ann and Dia Sheeta's apartment for one of their friends, a British lady living in Egypt with her daughter and son-in-law, an Arab gentleman and oil company executive. Mr. Saad Riad, a TWA sales manager, said that he could get me special-priced round-trip tickets for only $400 from New York to Cairo to bring my wife over if she would like. His wife was a charming talker and looker who was completely dumbfounded and unable to comprehend why I was drinking orange pop instead of scotch and soda. However, the fact that I was Virgo seemed to make it all clear to her. Mr. Riad's main concern was to help me understand why Arabs and Israelis don't get along. Before leaving the party, I promised to come back and help Ann with her roach problem.

One Sunday, Dr. Kamal, a fellow entomologist and associate of Isa's, brought me more literature on the brutal history and statistics of the Arab-Israeli conflict. I found myself increasingly ashamed of my country's policies that continued to support Israel in gradually increasing its boundaries in the name of security that made little sense in view of their relative military strength. It was sad to hear and see evidence of the suffering that Palestinian Arabs have endured in this conflict. It seemed that one of their greatest needs was to be able to publicize their plight to the world. My daily Bible reading had recently reached to the book of Joshua in which the brutality of the Israelites toward their enemies and the miracles performed in their behalf were food for thought. In view of the golden rule to "love your neighbor as yourself," I thought a possible conclusion to be drawn might be that God is showing that by making such greedy humans his chosen people, he cannot fail to include the rest of the human race as well, which Paul's mission to the Gentiles later confirms.

The remainder of the week was relatively uneventful. The monthlong Ramadan period of fasting from sunrise to sunset ended Thursday. We checked cane samples from the truck survey, finding little borer damage, which caused me to wonder again how random the sampling process had been. A cease-fire was agreed upon on Monday. But as soon as we again began to plan my survey trip to Upper Egypt, the fighting resumed on Tuesday, and our plans were again scrapped.

My Egyptian friends had become so deflated with Israel's success in the war that no one now mentioned it at all. I was disgusted with a situation in which I was spending most of my time in a chair at the office or at the Mayfair. Afternoon conversations with Tolstoy provided some diversion, but this was a far cry from what I had imagined this trip would be. I had now written four letters to my wife and children and still had received no mail. I was very homesick and becoming depressed.

In late October, I attended morning services at Saint Andrew's United Church, an old church at which fifty or sixty people were present that morning. After church, Mr. Shoucair, owner of the Mayfair, engaged me in conversation for a while to explain the agrarian reform, begun in Egypt about 1954 under Colonel Nasser following a revolution, by which the land was redistributed in parcels of about fifty acres to each qualified farmer. He admitted that poverty was still a predominant characteristic of the countryside but that some change had been made in the right direction. Mr. Shoucair later sought to give me a friendly verbal thrashing for the United States' position on the Middle East and for the general ignorance and lack of concern among average Americans about world affairs. I felt uncomfortable as an Arab sympathizer trying to explain U.S. policies on the basis of voter ignorance of human rights violations.

In early November, I mailed a fifth letter to Janice and gave two audio tapes to Dr. Rao to mail to her from India where he would be the following week. I spent an entire morning talking with entomologists in the Division of Biological Control at Giza and left feeling that I had stimulated some interest in potentially new areas of work for them by suggesting the use of insecticides as tools to demonstrate the value of beneficial insects. I began to think about a project report to be turned in at the conclusion of my mission. Daily walks to the UNDP office for mail continued to be fruitless.

Dr. Isa and Dr. Rao, with two other scientists and me, departed from Cairo by truck for Alexandria and a three-day visit to the experimental farm there to see flower initiation in sugar cane. We stayed one night in the Cecil Hotel on the Mediterranean coast. After learning that the experimental farm was off-limits at this time to foreigners, we drove along the coast from one end of Alexandria to the other and through the grounds of the Montaza Palace, a second Alexandrian summer residence of the Mohammed Ali dynasty rulers. We walked on the beach at a nearby resort where I dipped my hands in the Mediterranean. We visited the Ras el-Tin Palace, but could view it only from the front gates, and the Aquarium at the edge of the Eastern Harbour, but it was closed. Anti-aircraft crews were in evidence everywhere manning their guns. We stayed a second night at the Lido Hotel, where I was impressed by the accuracy of the water stream in the "tail-washing" commode (bidet). The Mayfair had similar commodes, but this was the best I had seen. Isa and I returned by train on Monday to Cairo where no mail had yet arrived from home. I wrote a sixth letter to Janice and a card to Mother that Dr. Rao took with him two days later to mail for me from India.

A truck returned from southern Egypt with forty-three cane samples, thereby finishing our planned truck survey of insect damage without my yet having put foot in a single field outside of Cairo. The following day, we finished examining the samples in which borer damage was so light that everyone was surprised. I was certain that the workers taking the samples were doing their best to show us the cleanest and nicest cane stalks they could find. Skirmishes between Israel and Syria had been reported the day before. But that day, Dr. Kissinger and President Sadat had talked, and diplomatic relations were restored between Egypt and the United States.

The following day, November 7, my first mail from home arrived—a letter from Janice and one from Mother. Dr. Stewart also told me that day that I should plan to give a two-hour seminar December 4 on the subject of Recent Trends in Entomology, which should be ready for duplication a week in advance. With gratitude for all my blessings, I sprayed the Stewart's home and the Sheta's apartment to control a horde of pesky German cockroaches. Three days later, I was busily writing an outline for my approaching seminar when Isa informed me that I was invited to address the Egyptian Entomological Society before leaving the country.

One day in mid-November, I began to feel sick and, the next day, to suffer with diarrhea. In spite of this, I spoke for more than an hour in the morning at a seminar for entomologists of the ministry on the subject of development of a pest management program. By noon, I was taking medicine to kill the strange bacteria in my gut, and all night long, I suffered with violent intestinal and stomach cramps that made sleep impossible. The next day, I arose feeling much improved, ate a light lunch, and spent the afternoon reading and organizing notes.

A travel permit was again issued for me to travel throughout the country south of Cairo, and this permit was not cancelled. We went by train in late November to Mallawi, about one-hundred-fifty kilometers distant, and made a second trip all the way to Aswan in the south later in December. I walked to the telephone office and paid for a call to be made to Janice at 7:30 AM the next day. When they called the Mayfair at 10:30 AM, I had given up and left but called them back and asked that they place the call again for Monday at the same time. When no call came again Monday morning, I returned to the telephone office to complain, upon which they promised to try again at 6:00 PM. This time, I got to talk to Janice for five whole minutes, during which I was given good news about the family but news of problems with the business. I wrote an eighth letter to Janice with cards to Nancy, Daniel, Mother, and Aunt Margaret, which were mailed the next day. I reviewed manuscripts for Isa, summarized truck survey data, longed for my wife and children, made a second payment on the ring for Janice, and visited the Brooke Hospital for Animals with Dr. Szydlo by invitation from Mrs. Blenman-Bull, an English lady at the Mayfair.

In my first trip to the south, Isa and I left Giza by train for El Minya, where we were met by an agricultural officer and workers for the sugar company at Abu Querqas. We examined several cane fields, where at the first stop I walked into the cane. Immediately, a company officer signaled that I shouldn't do that, that the labor would pull the samples, and that I should stand beside the field and wait for them to bring the canes out. The idea was that no "doctor" should have to walk in a dirty cane row to cut stalks and carry them to the headland. I suggested that each man collect a ten-stalk sample, including me. Reluctantly, he agreed. Upon examination, the stalks cut by the workers were practically free of any damage while my sample, selected without looking, had several damaged stalks. We held class there in the field, for the workers' benefit, on how to collect random cane samples to determine the amounts of borer damage present. After that, I felt better about the data we were collecting.

Later, we went to meet the director-general of the sugar factory, with whom we had lunch, before napping for two hours in the rest house. In the early evening, we met with company field technicians for a short talk and question-and-answer session, returning afterward to the rest house for dinner at about 10:00 PM. I thought that living as a guest in a sugar company compound was something like living in a country palace. Most of the next day was spent surveying cane fields in the area for insect damage. Besides sugar, there were molasses mills to which cane was hauled in twin packs hung across camels' backs. Crop residues were piled on top of buildings for fuel, and young children gathered fresh buffalo manure to be dried for fuel as well as fertilizer. Dinner was served at the rest house in midafternoon, and we napped afterward until 7:00 PM when a company official called to take us to a movie at the clubhouse before supper.

The next day, we left Abu Querqas by car for the Mallawi Research Station, arriving there in midmorning to be met by the director, who supplied us with helpers to examine sugar cane variety plots. Here we saw the most borer damage I had yet seen in Egypt. We lunched with the director in the station rest house, after which Isa napped as I summarized some of our data. Dr. Negm arrived with a graduate student from Asyut University to assist with the following day's work. We started the next morning with five people counting, Dr. Isa recording, and a dozen workers cutting and carrying samples while I supervised the sample selection process. Fieldwork was completed by midafternoon when lunch was served, after which I napped for a couple of hours before visiting with the station director and staff members at the clubhouse while Isa summarized data for the day in his notebook. We were served dinner of boiled eggs, white cheese, and yogurt before going to bed about midnight.

We were up early the next morning to catch a train to Cairo. I dined in the evening at the Mayfair with Dr. Szydlo, who brought me up-to-date on local news. I wrote a letter to Janice before reading my Bible and going to bed. The next day was spent summarizing survey data, eating oranges to ward off colds, and reading. The room was cold now both day and night, and there were no heaters in the bedrooms. I exercised vigorously, running in place and doing push-ups before going to bed.

Friday, November 30, started as the best day in several weeks when I received a letter each from Janice and Jan with pictures of the family and the August wedding. However, Janice's ten-page letter, in which she described her feelings of having been a neglected wife for twenty years while I sought one mountain after another to climb without having time or romance for her, hit me like a ton of bricks. She said that she had received all my letters and three tapes without one personal message "that no one could read but me, or a few personal words that no one else can hear but me," and she wondered if it could be that I just didn't feel that way about her anymore. After these initial charges, the remaining eight pages were newsy, loving, and reassuring. I spent the weekend preparing my lecture for next Tuesday, shopping for color slides of Egypt, talking politics with Mrs. Sidky and Ibrahim at the UNDP, shopping with the Stewarts, listening to news on Voice of America, and prayerfully reading my Bible while hurting and wondering what I could presently do about the situation at home.

I conferred with Dr. Stewart and Dr. Isa about the possibility of going home immediately to solve serious problems there, but there seemed to be no way out of my commitment at this time, and the red tape involved would not permit me to get there much sooner than I would be arriving there anyhow. Several days were spent at the office summarizing data, preparing a lecture, and reviewing our schedule for again visiting Upper (southern) Egypt. My two-hour lecture to experiment station personnel on the subject of Recent Trends in Entomology, with live translation provided, was accomplished December 4. That evening, I wrote more cards to family and fretted about Janice's last letter. Not only did her words about our relationship sting but hearing that Nancy cried every time she heard my voice on tape brought tears to my eyes as well.

One day, Isa and I finally got permission to visit the Cairo Museum and see some of the mummies of the pharaohs. On another day, we visited the pyramids of Giza, about eight miles from the Mayfair. Isa was smiling widely as the bus sped along when I asked him what was so funny. He said, "This will be my first time ever inside a pyramid. I used to pass them frequently and see them from a distance through the school bus window as a child but never before had the urge to take a closer look." The Giza pyramids are one of six pyramid groups spreading over some twenty miles on the eastern edge of the Libyan Desert plateau. There are three large pyramids in the Giza group, and each of the groups has large pyramids with smaller ones and tombs of nobles all around. The largest and oldest at Giza is the Great Pyramid or Pyramid of Cheops, completed around 2690 BC. I marveled at the thought that these structures were more than a thousand years old when Joseph and Moses were in Egypt. Approximately 2,300,000 blocks of stone, each weighing an average two-and-a-half tons, were used to erect the Great Pyramid, now four hundred fifty feet high after losing the top thirty-one feet to earthquakes, wanton destruction, and weathering. When first completed, it was covered with white limestone blocks of which now only a few remained at the base below the entrance.

With a guide to lead us, we entered the Great Pyramid at a point more than fifty feet above the base and walked about sixty feet to an ascending sort of gangway that was narrow and steep. This ascending passage was less than four feet high, but handrails and electric lighting had been installed to make travel easier. After crouching and stooping for more than a hundred feet, we reached a grand gallery over one hundred fifty feet long and nearly thirty feet high with more stairs. On the stairs with handrails, we climbed to the end of the gallery to reach another small passage, just over three feet high and about twenty feet long, which led to the king's burial chamber itself. We were breathless upon entering the chamber that was more than thirty feet long, nearly twenty feet high, about seventeen feet wide, and of black granite blocks. The granite sarcophagus, in the middle of the room, was empty when discovered many years ago, and the whereabouts of the mummy of the pharaoh and the lid that once covered it are unknown. Other passageways and chambers, such as the queen's chamber, were closed to tourists. There were no decorations or inscriptions seen inside the pyramid. Before the day was over, we also had visited the Great Sphinx and other ancient tombs and temples in the area, including a funeral parlor where there was a sunken stone tub in which mummies had been prepared for burial. This tub reminded me of our custom-made sunken tub, with ninety-degree angles and no curves, in our master bedroom at home. Now I had another reason not to like it.

Isa and I left Giza by train on our second trip to the south and arrived in Nag Hammadi in late afternoon when we were met by Mr. Kamel, who took us to the rest house at the largest sugar factory in Egypt. Although built in 1896, it now processed ten thousand tons of cane per day and handled about forty-eight thousand acres of crop per season. We had lunch there and napped for a couple of hours before dining with Mr. Kamel and other company personnel. Later, we watched TV until midnight. The next morning, we lost about two hours getting my passport registered with the local authorities but finally got into the fields around 11:00 AM. As always, we were provided with plenty of help who were more than anxious at first to show me the very best canes they could find in their samples. However, after some orientation, everything went well. We examined Isa's row-width experimental plots and several other fields before quitting time in midafternoon. After lunch and a shower, we napped a couple of hours before meeting the director general of the sugar factory, Mr. Hani El Zieny, who joined us for dinner and a long visit afterward.

The next day, Monday, December 10, we completed surveying fields before lunch and departed by auto for Qus, where we waited two hours for the police chief to finish his nap before registering my passport in his governorate. The fact that wartime rules were still in force was also evidenced at every bridge by manned anti-aircraft guns. Mr. Abdel Aziz, field manager for the Qus Factory, met us in the restaurant where we waited and accompanied us to the factory rest house where we dined, watched TV for a while, and went to bed.

The following day, we surveyed fields before returning to the factory and meeting assistant director general, Mr. Helme Boulas, who told us all about the history of the mill, its productive capacity, and projected growth. Following a late afternoon lunch, we departed by car for the Armant Factory near Luxor. We were met by a driver, who took us to a Nile River Crossing Boat on which we crossed the river to the west bank under a gorgeous full moon. I thought how much Janice would have been thrilled by this scene. We were taken to our rooms in the Aboud Pasha New Palace, surrounded by gardens, where we dined and visited with sugar company personnel before going to bed.

After breakfast the next morning, we met the director general of the Armant Factory, Mr. Osman Attiah, and field manager, Mr. Ahmed El Sissi. Mr. Ahmed accompanied us on our field survey, after which we dined with Mr. Attiah and other company personnel and walked in the palace gardens before going to our rooms to rest. That evening, we went to the clubhouse for a movie, Song of Norway, before returning to the palace for a late supper. I went to sleep reading a Field Guide to Egypt in preparation for the next day's trip to the ancient sites.

Mr. Milad, the company public relations officer, took Isa and me to see the ancient sites at Luxor and the ancient city of Thebes. We visited the tombs of Tutankhamun; Seti I; Amenophis II; the tomb of the chief artist, Sinnegem of the Nineteenth Dynasty near the temple of Deir al-Madinah; and a temple (palace) of Queen Hatshepsut. Some have speculated that Hatshepsut may have been the pharaoh's daughter who adopted Moses. With this thought in mind, I stood in the front yard of the palace trying to picture little Moses running about playing on that bare ground without a blade of grass. We were guided on the tour by a government antiquities inspector, Mr. Sabbahi. Our next stop was an alabaster factory where I bought a couple of souvenirs and stood at the feet of the Colossi of Memnon for Isa to take my picture. We crossed the Nile to the east bank and joined Dr. Sayed, expert on Egyptology, who accompanied us to the Temple of Karnak, where we paused at sunset for a cool glass of lemon pop at a single table with three chairs beside the Sacred Lake. Dr. Sayed read hieroglyphics on the temple wall, including accounts of wars with the Hivites, Hittites, and Amorites long before the Israelites with Joshua had entered the Land of Canaan, as if he were reading a newspaper. There were no tourists present at any of these sights due to the war. Anti-aircraft guns were manned at every bridge we crossed and every hamlet we visited. All had been arranged for my entertainment by the sugar companies in cooperation with the government. We returned to the palace for dinner, again by boat and well after dark. Although I was saddened by news from home and suffering from a cold, this day was a highlight of my visit to Egypt in which my thirst for seeing ancient sites was well-watered.

    The next day, we left the Armant Factory by car for the Matana research station where we went with the station director, Mr. Farrad, directly to the field to sample and examine variety plots. I became very weak due to my cold that had worsened to the point that I was forced to leave the field and go to bed. I felt better after sleeping for about six hours. After eating and visiting the juice-analysis laboratory and the clubhouse, we visited with Mr. Farrad and his staff until 10:00 PM. The next morning, we toured the station as Mr. Farrad explained all the experimental work going on there and then departed in a UNDP truck for Esna and the Edfu Sugar Company.

    A company car met us in Esna and took us to the Edfu Factory. However, we stopped on the way to photograph a large area, walled by ancient sun-dried brick, which I was told had been used by Joseph, son of Jacob, over twenty-six hundred years ago to store grain in preparation for the seven lean years. We lunched at the factory rest house with Mr. Kamel, director general of the factory, and Mr. Ayad, the field manager. We slept for a couple of hours before visiting with these gentlemen for another several hours during which our conversation centered on religion. Mr. Kamel sought to explain Islam to me while politely expressing interest in some of the similarities between his faith and mine. Dinner was served about 9:00 PM hosted by Mr. Ayad, as Mr. Kamel was excused to meet other visitors. The next day, which was Sunday, we finished surveying fields around Edfu and visited the Temple of Horus, the best preserved and second largest of all the ancient temples of Egypt. After lunch, we visited in Mr. Kamel's home meeting his family before leaving for Kom Ombo.

Upon arrival at the factory rest house in Kom Ombo, we were greeted by Mr. Ismail Sabry, general director of all the sugar companies of Egypt, and other officials with whom we dined and who were also staying at the rest house this night. Mr. Sabry asked for my recommendations, now that I had seen so much of the country. I told him that I was still in the process of formulating recommendations for my final report, but I believed that an integrated pest management (IPM) program, with the occasional use of pesticides as a last resort in some places and also as an additional research tool for evaluating the effects of beneficial insects, should be an important part of an appropriate solution to his insect problems. The next day, we surveyed cane fields in the vicinity and southward until 2:00 PM, when we returned for lunch and an afternoon nap. At the rest house, we attended a movie entitled How to Save a Marriage and Destroy a Life, an interesting title about which I remember nothing else. After the movie, Isa and I and a Dr. Ibrahim Mansown played dominoes til dinner at 10:00 PM.

This day, we were driven by car to Aswan to see the Aswan High Dam before lunch at the Cataract Hotel. Aswan is about one hundred fifty kilometers, as the crow flies, north of the Sudan border. However, the Cairo to Aswan railroad ends there and goes no farther south. After lunch, we took a sailboat to the west bank of the Nile to see the tomb of Aga Khan, who died in 1957 while spiritual leader of a Shi'ite sect based principally in India but with followers around the world. His mausoleum is an elegant pink granite structure on top of a hill with an excellent view, which includes his white villa below where his wife lived until her death in 2000. He is said to have been a very large man and extremely wealthy, so much so that on his birthday in 1945, he was weighed in diamonds that he then distributed to his followers. We then sailed to Kitchener's Island, a botanical garden filled with exotic plants and trees imported from all over the world. The island had been given to Lord Kitchener for his campaigns in the Sudan, and he had moved there and created this garden, which the Egyptian government now operated as a popular tourist attraction. At sunset, we sailed past the rock tomb graves of the nobles cut into the face of the cliffs on the west bank during the Old and Middle Kingdom periods (2700-1600 BC) before returning to Aswan on the east bank and a car ride back to Kom Ombo for dinner, dominoes, and bed.

On the final day of this trip to the south, Wednesday, December 19, I was introduced to several classes of students at a vocational agricultural school in Kom Ombo. They showed me around their laboratories, in one of which the art of making white cheese was demonstrated for my benefit. We returned to the rest house for lunch at 2:30 PM, then packed and departed for the airport where we boarded a small Russian-made airplane in which we arrived in Cairo at 6:40 PM and the Mayfair by taxi at 8:00 PM. I ate dinner, unpacked, and went to bed to keep warm and nurse my cold. It was noticeably colder in Cairo than it had been in Aswan, where the weather had been delightful.

My lasting impressions of rural Egypt are of masses of people dressed in coarse white clothing, reminiscent of biblical times, and engaged in lifelong struggles for survival. I felt that I could relate to them better than many of their own countrymen because of my own childhood during the Great Depression. Although I had never plowed with an ox or camel, I had plowed with a mule, and while I had never worn white robes, I had worn white trousers and shirts made from flour sacks. I had seen food and services exchanged by barter in place of cash. I had drawn water from a well with a bucket on a chain to drink and bathe; they drew water for their gardens from an irrigation ditch by means of a bucket attached to the end of a long pole. After a long day in the fields and a bath in the irrigation ditch, they gathered at the community center to smoke and watch a single black and white TV set on which they saw life in Europe and America. I wondered for how long they could be content with their lives after seeing how more affluent cultures were living. The Aswan High Dam was supposed to have raised their standard of living, but it had increased the prevalence of schistosomaisis by creating more habitats for the snail hosts of the blood flukes that slowly weakened and killed masses of the people. Many were becoming infected while bathing in the irrigation ditches. Control of flukes could be carried out by educating people to dispose of body wastes hygienically, but this is difficult with poor people living under primitive conditions.

My first day back in Cairo was spent at the office where Bill Awadallah worked at the calculator computing percentages and averages from our field data collected during the last eleven days. During my absence, letters had arrived from my brother Needham, Aunt Chloe, and Mother. At dinner, I wished Dr. Szydlo a merry Christmas as he was leaving early the next day for ten days in Poland with his wife and children. That evening, I read ten chapters in II Chronicles, ate oranges, exercised, and wondered what was going on at home, as I did every day. But that evening, it was on my mind more than usual. I was envious of Bogdan Szydlo. The next day, I made air reservations for my return trip home to depart Cairo January 8, 1974, and to arrive in New Orleans the day after at 9:00 PM after a stop in Rome. Isa wanted me to get a tailor-made suit of Egyptian wool, but when we priced it, I decided that it was too expensive. I spent the evening reading my Bible and studying data from the field survey.

Saturday morning was spent at the office, and in the afternoon, I shopped for Christmas cards and a cake for Marion Stewart, who had suffered a heart attack while we were in Upper Egypt. I wrote a letter to Janice telling her about the trip and giving her the particulars of my schedule for returning home. Sunday, I worked at the office on survey data in the morning and took a walk after lunch through the grounds of the Gezira Club. In the evening, I read scripture, worked with survey data, and prepared for a talk on Monday to a special meeting of the Egyptian Entomological Society. Before the meeting, I lunched with Isa and Dr. Hafez and went to Isa's home for oranges. We took a taxi from his house to the meeting place where tea and cookies were being served when we arrived in late afternoon. I addressed the society for more than a hour on a variety of subjects, including the use of insecticides in integrated pest management programs, how such a program had been developed in Louisiana, and how such programs, with appropriate variations, should be applicable anywhere.

On Christmas Day, Isa still was determined to get me into an Egyptian wool suit before I left. He insisted on going with me to buy wool and to have me fitted by his tailor. I caught a taxi from the tailor to Sheeta's apartment, getting there just in time to go with them to the Stewarts for a Christmas meal and to visit with these families and several of their friends. The next morning, I started in earnest writing my report on the "Present Status of Sugar Cane Pests and Recommendations for Development of a Program for Their Management" based on the work of W. H. Long, consultant in sugar cane pest management, as part of the project on "Improvement of Field Crops Productivity in the Arab Republic of Egypt" by the Food and Agriculture Organization of the United Nations, Rome, 1974. In the evening, Mr. and Mrs. Shoucair took me to dinner at their private club with another friend of theirs who was visiting from Canada.

There was a letter waiting for me the following day at the UNDP from my eighteen-year-old daughter, Jan, a sweet letter from a loving daughter and a delightful break from report writing. The following day, I talked by phone to Janice, Nancy, and Dan in the late afternoon in anticipation of my arrival home in ten days. Two friends from Asyut University came to visit later. We dined at the Mayfair and later went to a movie. The next six days were spent working on my report.

During this time, I met an interesting young German lady, Ms. Erni Leppert, at lunch in the Mayfair. She was a press agent for the German government, had a favorite hobby of collecting old books and a most infectious laugh. We had a friendly debate for more than an hour on the pros and cons of the women's liberation movement and went from that to capitalism versus socialism. I was the drab and conservative capitalist and she the engaging, kind-hearted, liberal socialist. She argued that my capitalistic ideas were not consistent with my Christian faith, etc. It was an enjoyable diversion from report writing.

After working several hours at the office in the morning, I returned to the Mayfair in late afternoon to find Erni Leppert already having dinner when I passed by the dining room. I intruded upon her privacy to her apparent pleasure. She proudly showed me her shopping triumphs from the Khan el-Khalili Bazaar and asked me to go with her again tomorrow. I declined as I had an appointment with the Stewarts at that time but agreed to go with her the following day. She said, "I'll be leaving for Berlin early Thursday morning. We don't have much time." I said, "I know how that is. I'm still writing my final report to FAO, and

I've got to get on it tonight, tomorrow, and the next day too." I spent the rest of the evening working on the report.

Monday morning and early afternoon was spent at the office writing before going to the Mayfair for something to eat. Erni was sunning on the balcony with lemonade, so I joined her there with my lunch, and we chatted until the Stewarts came to pick me up. They were going to Khan el-Khalili to order a set of silver spoons, and I was going there to make the final payment and get Janice's pyramid ring, which was now completed. On the following day, the final report was still my major activity and concern. But after a midafternoon lunch at the Mayfair, I accompanied Ernie, with two other Mayfair residents, to the Khan el-Khalili Bazaar. We returned about 9:00 PM for dinner and sat and talked afterward until midnight.

I worked on my report most of the next day, except for a trip to the barber. I was having dinner at the Mayfair when Erni showed up and joined me. After dinner, we talked until 2:00 AM. The final report was finished the next afternoon, and I packed my books to be sent home via airfreight. Erni and I visited again after dinner that evening in the Mayfair sitting room until after midnight. She was a very nice person with an interest in everything under the sun. She wasn't particularly religious but was curious about my faith. She wanted to know more about it, and I answered her questions as well as I could. We shook hands and hugged before parting, and her flight left for Berlin three and a half hours later.

In my final report, I described results from field surveys that described the distribution of crop damage from two species of cane borers, the purple-lined borer and the pink borer, and the wide distribution of the pink mealybug. I recommended that breeding for borer resistance be incorporated into the sugar cane breeding program, that efforts be continued to establish new parasites of these pests, and that some immediate relief from borer damage might be available through row spacing, irrigation control, and leaf removal or some combination of these actions as indicated by recent experiments. I also recommended that insecticides could be profitably and safely used in a well-planned and integrated pest management program, but should always be considered as a last resort. I made specific recommendations for development of such a program. Other recommendations called for strengthening research staffs at the experiment stations, improved library facilities, and more interdisciplinary cooperation in research.

After reviewing two short manuscripts for entomology staff members Friday morning, I took my final report to Mrs. Bonfanti for typing at the UNDP. The next day, Isa and I lunched together before going to his house to spend the afternoon with his wife and children and his ministry associate and neighbor, Dr. Hafez. When no taxi could be found later, I walked back to the Mayfair. On Sunday, I took Don and Bonnie Stewart to lunch at Andrea's Restaurant near the pyramids and took a chicken dinner back to their home for Marion, who was still convalescing. That evening, the plant breeder, Dr. Rao, served refreshments at his home for Mr. and Mrs. Sidky, the Stewarts, and me. The following day, I was alone at the Mayfair for lunch when the Stewarts came by with letters to be mailed in the States and to join me for coffee on the veranda. Dr Allam came in the early evening to dine with me at the Mayfair, stayed afterward to help me pack, and then went with me to some nearby shops in Zamalek to help bargain for a few souvenirs with my remaining currency. I finished packing later and went to bed before midnight.

Tuesday morning, I was up early to meet Isa, Bill, and the UNDP driver in front of the Mayfair at 4:00 AM. They accompanied me to the airport, where we parted with the traditional hugs and kisses of Middle East cultures. It was a sad moment, realizing that we probably would never see each other again. I went through Customs in less than an hour and boarded TWA Flight

901 for Rome, arriving there at 8:15 AM. After checking my big bag at the airport, I took a taxi to the FAO building to turn in my report and for debriefing. This included verbal reports to Dr. Buyckx and Dr. Ahmed, a visit to the finance office about final pay and expenses, and the first part of a final physical exam for which time ran out that afternoon. A reservation was made for me at the nearby Hotel S. Anselmo from which I called Mrs. Bonfanti's brother and left a package for him in the hotel lobby. The next morning, I was up early for shave, shower, and orange juice before completing my physical exam. I would be home or at least in Louisiana before midnight that day.

My lasting impression of the Egyptians, with whom I had socialized and worked for four months, was that of an intelligent, educated, religious, honorable, sensitive, caring, and loving people. There seemed no limit to the things they would do for friends and loved ones. The Arabs that I met liked Americans but did not like our government's one-sided support of Israel. Except in the cities of Alexandria and Cairo, the latter with a population of several millions, there was little evidence of a middle-class culture. Although this is not a characteristic of governments dedicated to the promotion of freedom for its people, small positive changes had been made. Meaningful democracy evolves slowly as more people become enlightened, educated voters.

Several months later, I wrote a letter to the editors of three Louisiana newspapers, condemning U.S. policies that blindly supported Zionist ambitions to increase Israel's boundaries. The Baton Rouge and New Orleans papers never acknowledged receipt of my letter. Only my home town paper, the *Daily Comet* of Thibodaux, Louisiana, published it on June 11, 1974. This said tons, I believed, about the Zionist influence on our free press.

12

MIDLIFE CRISIS (1974)

Janice greeted me at the New Orleans airport at 11:30 PM, January 9, 1974. She was beautiful, but had a troubled look. When I hugged and kissed her, I could tell that something was very wrong. As it was already late, we checked into the nearby Ramada Inn. She told me that she had come to believe that I didn't love her and that proof of the matter was in the fact that I could leave her for so long and with so many problems and responsibilities to deal with alone and without my help. I was shell-shocked by the news, physically exhausted from the trip, and emotionally incapable of reacting. Neither of us slept a wink but talked the rest of the night. I gave her the pyramid diamond ring from Yazejian's in Cairo to replace the very small diamond that her student husband had given her twenty years earlier. I claimed that life had not been exactly a picnic for me during this time and that I thought we both had sacrificed for the good of all. The more we talked, the more apparent it became to both of us that we had become emotionally separated in a deeper and more serious manner than ever before. The following weeks and months were terrible. We wanted to mend things and make them like they used to be, as imperfect as that had been. Our emotions and feelings for each other waxed and waned daily like a yoyo on a string. We tried and cried and prayed and tried some more until we both were approaching nervous breakdowns, or so we felt.

On recommendation from my brother, Needham, we scheduled an appointment with Dr. John Reckless, a psychotherapist of Durham, North Carolina. His spoken diagnosis was that Janice was as "normal as homemade apple pie" and had endured twenty years of emotional hardship living with me, a man who had been raised in such a strict and puritanical manner that many women might have difficulty adjusting to life with him. His word for me was that, no matter how hard I might try, I would never be able to measure up, in her mind, to what a husband should be. His words to us in a letter, written the day after our departure, were that he hoped the explanation he had given us had "brought together the various elements of our situation." He urged us to seek further professional counsel and to "work to improve within our present relationship the situation that exists rather than to dissolve the years that we had spent together." He urged us to come back to North Carolina to "take advantage of a two-week experience at his clinic" and to understand that the first step should be "to completely understand the full dimensions of the problem and then to go from there to treatment." I believed that we could not afford any more of his services and that perseverance and hard work would solve all problems. We began reading together in the evenings The Art of Loving and attending dance classes. We both had ingrained religious beliefs that marriage was forever. For months, we alternated between periods of earnestly trying to be more attentive and appreciative of each other and periods of anger, but nothing seemed to work. It was hard to think about the impact of all of this insanity on our young children, Nancy and Daniel; their emotions; their faith; their concepts of family; and on our daughter Jan and her family. Finally, we decided that only separation would sustain our sanity and health. And so on a morning in October 1974, Janice and the children drove away in the family station wagon headed for North Carolina while Nancy and Dan waved goodbye to their daddy and clutched homemade rag doll, Halloween ghosts, which had hung from a tree in the front yard. Janice's eyes were full of tears as were my own. I was now forty-six years old and had been married for twenty years. This was our second separation, the first having occurred briefly for a few months one year after our marriage while I was pursuing graduate work toward the PhD degree at Iowa State University. Within a short time, I became involved in an affair, which lasted several months, with an unmarried female beautician in her late twenties.

Weeks later during Christmas week, I flew to Raleigh to visit my children and to see Janice. On my last evening there, she entered my bedroom to check on me and say good night but didn't leave until sunrise. Hours later, we all kissed and waved goodbye at the airport, and I returned to Louisiana a confused human being to pack my bags for a four-month assignment (January to May 1975) in Brazil with the International Atomic Energy Agency (IAEA) of the United Nations.

Upon my return from Brazil, I appeared by invitation in June 1975 before the New Orleans High Council of the LDS Church, which I was not obligated to do but at which I wanted to make several statements: (1) I respected the men present and had enjoyed working with them over the years. (2) However, I could no longer believe that instructions from Salt Lake City regarding church activity were invariably superior to contrasting ideas generated by grassroots leaders and members on the scene. (3) While I had great respect and appreciation for many teachings of the church, I did not believe them all. (4) I was still unhappy for being censured in the late 1960s by the presiding authorities for teaching Darwinian evolution. (5) I believed that my recent adulterous relationship had been beneficial to my mental and physical health. (6) I no longer believed that LDS priesthood bearers had any more inspiration or authority to act for God than did other honest, sincere, and prayerful Christians. Therefore, I felt that I had no business remaining in the church.

Privately, I had additional feelings that I did not express. I was relieved to be out of this church organization that had long required more of my time and energy than I believed was reasonable and one that, it seemed to me, had played a role in wrecking my family. However, I also recognized that, after fifteen years of active church membership, involving preparation and administering sacrament; teaching Sunday school and priesthood classes; preaching sermons; ministering to the sick; conducting various kinds of meetings, funerals, and even a wedding; interviewing for calls or temple recommends; and a host of other activities; I had come a long way from the boy who could not deliver a five-minute talk in ninth grade and the LSU entomologist who downed six martinis in less than six minutes in order to speak to a full auditorium at the annual Contact Committee meeting of the American Sugar Cane League. I knew that, in some respects at least, my active years as an LDS church member had fulfilled the Book of Mormon promise that "if they humble themselves before me, and have faith in me, then I will make weak things become strong" (Ether 12:27).

My friend and stake president, Dr. Melvin L. Gruwell, who was also dean of the College of Education at Tulane University, responded to me verbally and in writing that included the following statements and thoughts: "I can well understand your feelings and present concerns. I again repeat that you have always been honest with me. I do appreciate this. It is with sincere regret that I inform you that the action taken by the court was that you be excommunicated. I sincerely hope that you and Janice are able to reconcile your differences. I know that you can. Her remaining close to the church and getting away where she could see things objectively have had strong influence on what exists today. I realize that your testimony has been shattered by the sequence of events. But I also know you as a man of tremendous inner strength. Again, my prayers are with you and Janice, knowing that you can put it back together and eventually have the full blessings of the gospel restored."

In late 1974, I had been encouraged by a few persons to put my name in the pot for an upcoming vacancy as dean of the college of life sciences and technology. While in Brazil in early 1975, I sent my letter of application for the position. In spite of poor past relations with some of the administration, I did so and was passed over, not surprisingly, in favor of a more politically savvy candidate.

13

BRAZILIAN MISSION (1975-76)

In the fall of 1974, I received correspondence from an old acquaintance and fellow student from Iowa State University, Dr. Don Lindquist, and then head of the Insect & Pest Control Section of the International Atomic Energy Agency (IAEA) in Vienna, Austria. Don inquired about my possible interest in an eighteen-month assignment to Brazil that would emphasize field ecological studies to support the application of nuclear techniques in entomology there. If I accepted the job, I would be appointed as an expert in entomology working for the IAEA as part of a United Nations Development Program (UNDP) project, directed by a UNDP project manager. The assignment could be served in one or two tours of duty with some flexibility in the duration of each tour. The duty station would be at the Center for Nuclear Energy in Agriculture (CENA), located in Piracicaba, about an hour's drive west of Sao Paulo. CENA is a complex of laboratories and offices that make up a part of the agricultural college of the University of Sao Paulo. He was looking for an entomologist with several years of experience in insect field ecology, preferably experienced with either sugarcane borer (SCB) or medfly, and also with training or experience in radioisotopes. The job description further indicated that the work should be pest management oriented and should develop information on pest population dynamics, insect field behavior, and related subjects.

Further correspondence from Don indicated that one of the great needs of the entomology group there was for them to take off their white coats long enough to get out of the laboratory and into the fields to see what the problems were really like. My past record and experience suggested that I was well-qualified for the job. I wrote Don that I was definitely interested, that my family situation was uncertain as my wife and I were presently separated, and that my growing business would suffer at this time without my presence in Louisiana during late spring and summer. After further correspondence, I accepted his offer and proposed to break the mission into two tours, the first being from January to May 1975. In this way, I could get oriented and get some work started there before coming back to Louisiana for the busy summer consulting period. The IAEA agreed to my proposal and my employee, Charles Taylor, with his wife and two children, moved into our home rent free to care for it in my absence.

I flew to Brazil in the first week of January 1975 arriving in Sao Paulo after a twelve-hour flight and at CENA in Piracicaba an hour later in midmorning in a van driven by a "motorista" who met me at the airport. This driver spoke no English, and I spoke no Portuguese, but he was very friendly and talked with both mouth and hands all the way to Piracicaba. At CENA, I first met the UNDP office secretary, Dona Diva Athie, who warmly welcomed me and said that she had already reserved a room for me at the Hotel Esplanada. She introduced me to the project manager, Dr. Knut Mikaelsen, a large Swede, who also welcomed me and described the project there that, in addition to me, included Dr. Stig Blixt, another Swede and plant breeder also staying at the Hotel Esplanada, a soil scientist from Canada who was there with his family, and several other agricultural scientists from various places who were there or had been there and would be back.

He then took me to meet the head of CENA's entomology department, Dr. Fritz Wiendl, a short light-complexioned man of German origin with large thick glasses and an obvious problem with psoriasis. Fritz also had faculty rank in the department of entomology of the nearby Escola Superior de Agricultura "Luiz de Queiroz," da Universidade de Sao Paulo (ESALQ). Fritz was polite, but I immediately sensed reservation on his part, which I would understand better in the not-too-distant future. Fritz introduced me to Julio Marcos Melges Walder, a young doctoral candidate with whom I would be working most directly. Julio showed me around the laboratory and introduced me to the technicians Valdemar Tornisielo, Jose Peruca, and Valter Arthur.

I was assigned a desk, conveniently close to Julio with whom I would be working most directly, in a room shared by graduate degree candidates and technicians and adjacent to Fritz's office. Fritz's interests were focused primarily on laboratory studies determining the amounts of gamma radiation required to sterilize or kill various species of stored grain pests. This was important work, but I believe that he resented or felt threatened by any attempt to shift the focus of the entomology department toward more ecological field studies. In retrospect, I have wondered how much he knew about the purpose of my mission before I arrived. His attitude and that of some other entomologists there seemed to be that the important facts about field biology of insect pests were generally known and that the greatest needs were for the application of sophisticated scientific equipment to the solution of research problems in the laboratory. I believed that some of the basis for this kind of thinking was a natural aversion to physical labor by those with graduate degrees. Our different perspectives did not bode well for development of a close working relationship, or of any other kind, between Fritz and me. He seemed to suffer from an exaggerated professional pride associated with a mixture of national pride and disdain for the fact that an unassuming "americano" with muddy boots should be designated as an expert and assigned by the IAEA to his department at CENA. These circumstances placed Julio in the awkward position of having to work with me without displeasing or offending his graduate committee chairman, Dr. Fritz Wiendl. Julio would do a pretty good job of walking that tight rope.

The IAEA was still greatly interested at this time in finding more applications for the sterile male technique in insect pest control following the initial spectacular success with this approach in the eradication of screwworms from the state of Florida. This technique had required the artificial laboratory rearing of millions of sterile male screwworm flies that were released from airplanes in sufficient amounts to outnumber the natural female fly population by a ratio of ten to one. As the female flies mated only once, most of them laid nonviable eggs. After several such releases, the fly population was reduced to extinction. Success of the program had depended upon a detailed knowledge of population dynamics of the screwworm fly and other details of its biology, such as the average number of female flies present per acre, the time of year at which minimum fly populations occurred, sex ratios of natural field populations throughout the year, and how often the females mated. The entomologists at CENA and at the local agricultural college had no idea how many SCB moths were present per hectare, in what month of the year they were most abundant, what the natural sex ratio was, whether or not it was constant, and neither did I. There was only one way to find out, and it could not be done in the laboratory alone.

With a road map of Sao Paulo State before us, Julio and I divided the state into four regions for the purpose of surveying sugar cane fields. In each region, fifteen or twenty fields, five kilometers apart, would be visited monthly. At each field and visit, ten randomly selected stalks would be examined and dissected and the number and location of all SCB stages recorded. All live borers would be taken back to the laboratory and reared to adulthood to determine amounts of parasitism and sex ratios. This data would be collected for a twelve-month period from January to December 1975. We began this survey the third week of January with Julio and me and the three technicians in a personnel carrier accompanied by a motorista, surveying one region daily and usually completing the four regions in four days, weather permitting. Although time-consuming, this activity became routine after a couple of months during which a considerable amount of quantitative data on SCB biology was being collected. I presented a seminar on the Development of a Pest Management Program for Sugar Cane in Louisiana before an audience of faculty and students and as a part of CENA's technical and scientific seminar series.

During the next few days, weeks, and months, I was introduced to the director of CENA, Dr. Admar Cervellini, the other four entomology faculty members at the agricultural college, to scientists at the Copersucar sugar cane experiment stations at Piracicaba and at Sertaozinho, the Instituto Biologico at Sao Paulo, the Instituto Agronomico de Campinas at Campinas, the national sugar cane quarantine station at Maceio, the nuclear energy center of the federal university of Pernambuco at Receife, and Planalsucar's south central sugar cane experiment station at Araras.

At my first and only visit with Dr. Domingos Gallo, an elderly gentleman and head of the entomology department at ESALQ, he gave me a recently published thick hardbound "Manual de Entomologia," signed by himself and three other coauthors and members of the departmental faculty. He advised me that SCB control had been adequately addressed for a long time by releasing natural parasites from laboratories scattered throughout the country. When I mentioned our recent discovery in Louisiana of the importance of ants in this role, he affirmed that they were not important in this way in Brazil. He also warned me that there was no need for insecticide use in SCB control here. Dr. Neto, one of two Japanese members of the entomology faculty, was in charge of departmental seminars and invited me to present a seminar on the subject of integrated pest management that I agreed to do at a future date.

There was plenty of time between survey trips to study, plan, visit sites of interest, and socialize. Julio and his wife, Marcia, practically adopted me. I very much appreciated and benefited from their warm hospitality. Indeed, I almost felt like a member of their family, dining frequently in their home, and playing with their baby son, Mauricio. They took me for a weekend visit to Marcia's parents' farm, Cascada (waterfall), several hours' drive from Piracicaba near the town of Dois Corregos (two brooks). There were social events periodically at the home of Dr. and Mrs. Cervalini, and I was invited to dinner and to parties in the homes of the other IAEA experts. Dr. Stig Blixt and I became good friends, both rooming at the Esplanada Hotel where we waited each morning by the curb for our ride to work. Stig was a great storyteller with a never-ending repertoire of jokes that I enjoyed over a significant number of bottles of 8 percent beer (Brahma or Antarctica). Another stronger and favorite drink was the caiperinha, a mixture of strong cane alcohol with lemon juice and sugar. This appeared to be the most popular mixed drink sold in the bars where workers congregated after a hard day's work and while waiting for buses to take them home. One couldn't consume many of these and remain functional.

One day at the office, I expressed the desire to see what nightclub life in Sao Paulo was like. Before my four-month tour was up, Diva and I had dined and danced in Sao Paulo, a city of more than ten million people. Diva was an extraordinary, sophisticated, and well-educated lady from Brazil's upper class. Her brother was a retired attorney of considerable means in Sao Paulo. She spoke fluently and wrote equally well in several languages. She cared for the foreign experts, who were always arriving and leaving from CENA, like a mother hen cares for her chicks—helping them find housing, tending to all kinds of governmental red tape, typing their reports, and being genuinely concerned about their welfare. She had that unusual ability to make everyone feel special, and I was no exception. While CENA project managers came and went, Dona Diva Athie was really managing and pulling many of the strings herself. She became a dear friend with whom I corresponded at least annually after leaving Brazil until her death in 1999. In her last letter to me (1998), she described two books she recently had been translating from English to Portuguese, one entitled "Career Power!: 12 Winning Habits to Get You from Where You Are to Where You Want to Be" and a second on "microchemistry" (a college textbook from the U.S.), before her computer containing all her work was stolen. And finally, she wrote "It is good to learn that the whole family is fine and that you are doing exercises to stay young. I wish I could do the same. My best wishes and kindest regards to all of you. D. Athie." I sadly learned of Diva's death from cancer in 2004 while writing this chapter.

Family Reunited. April 30, 1975, I bid my new friends goodbye for five months and returned to Louisiana via North Carolina to see Janice and the children and to tend to business with my farmer customers for the summer. I spent the first ten days of May with Janice and the children in North Carolina. We were happy to be together again, and the children were ecstatically so. We decided, for their sakes and for the twenty-two years we had already invested in each other, to put our family together again in spite of all past events. Janice drove from North Carolina to Louisiana to spend ten days with me during the latter part of that month while leaving the children with Jan. This was more like a honeymoon than anything we had known before. We spent three days and two nights in New Orleans. We dined out, shopped, danced, and rekindled our love for one another.

On June 8, 1975, two rental moving vans left Raleigh, North Carolina, one loaded with Janice's household goods and the other with Jan's. I drove one and my son-in-law, Dennis, the other until we arrived in Thibodaux late the next afternoon. Charles Taylor with wife and two children had just moved out as Dennis and Jan with baby Holly moved into the second floor of our home on the bayou. We occupied the remainder of the house until October, when I would return to Brazil with Janice and our two young children. Dennis went to work immediately for Long Pest Control as a service man controlling crawling insects in homes and businesses. Janice and the children again attended LDS church services at the Thibodaux branch. However, I was too busy with farmers, weekends included, to think about church now. Besides fieldwork and regular office responsibilities, there was a lot of correspondence required in preparation for the second Brazilian mission. Charles Taylor was a Nicholls graduate who had taken my entomology course and gone to work for me after graduation. He had become certified as an agricultural consultant in entomology and also in termite, household, and horticultural pest control. The summer was a busy one in which I sought to assure our farm customers that they would be in good hands with Charlie doing all the field scouting next year while I would be in Brazil. Our office was in a rented building on Himalaya Street, and all office work was done by me or Charlie, with a CPA friend and neighbor coming in once a week to keep up with billing and financial records.

After a twelve-hour flight from Miami, we arrived in Brasilia, early in the morning of October 1, 1975, for briefing with government and IAEA officials. We bought an auto tour of the city with an English-speaking motorista and flew the next day to Sao Paulo, arriving there in midafternoon. The weather was stormy around Sao Paulo, and it seemed that frequent lightning strikes were just beyond our wing tips. Dan and Nancy enjoyed the excitement, but Janice and I were worried and happy to touch down, upon which the landing was as rough as the flight. Diva, Fritz, and the friendly motorista met us at the airport. We stopped for dinner in Campinas on the way to Piracicaba, a city with population over one hundred thousand at that time. We arrived after dark to check in at the Hotel Esplanada. After a brief appearance at the office the next morning, I returned to the hotel to walk around the central plaza and business area of town with Janice and the kids to help them get their bearings and a feeling for the new environment.

During this month, with Diva's help, we located and rented a house at 1123 Rua Edu Chaves where we moved after living for three weeks in the hotel. Dr. Admar Cervalini and his wife, Cida, entertained us at their home several blocks away. Marcia and Julio invited us to dinner several times at their apartment. The ladies visited Janice at our house. Janice was introduced to my American friends from the Methodist Church, Milt and Diana Lind and Bill and Marge Blockly. We had dinner in their homes, and they informed us about the schools and helped us to get situated. Instead of placing Nancy and Daniel in an English school one hour away in Campinas, we entered them into local schools where no one, including the teachers, spoke English. Nancy entered third grade at Dom Bosco School, and Daniel began kindergarten at Cigahinya. We also sent Nancy to a Brazilian lady for daily tutoring in Portuguese each morning. Two different buses picked up the children during noon hour to take them to their respective schools. We soon realized that the ringing of the bell on the "pao de azucar" (sugar bread) wagon was a sure sign that the school buses would be here momentarily. Daniel and Nancy commonly waited by the curb daily with a couple of cruzeiros each to buy their sweetbread.

The children were very frustrated for six or eight weeks but, after three months, were playing and talking easily with others in the neighborhood. Daniel became a sort of leader among the young boys on our street, who played soccer in the streets year-round, and he was "drafted" by a young men's soccer team to attend their home games as the team mascot. After about eight months, Dan was surprised one evening to find that he could explain something to me better in Portuguese than in English. Nancy also quickly became so comfortable with the language that we came to depend on her to help us understand waiters and sales clerks.

After we had been in our house a few days, we hired a highly recommended maid, nicknamed Tuca (Cleide Ricardo Teixeira). Tuca was in her early twenties, had never finished school, had two young children, and lived with her parents. She had worked with foreigners before and was happy to find employment with us. She worked ten to twelve hours daily, or more if needed, and six days weekly with Sundays off. She sometimes let herself into the house in the mornings to start breakfast, the first of two meals she occasionally prepared, and left in the evenings after washing and cleaning indoors and out. She always had a big smile, never lost her temper, and had the friendliest and most positive personality I have ever encountered. She talked incessantly, which drove Janice to distraction for a while. But after a time, Janice also began to talk to Tuca. When Tuca had been with us for about six months, I was conversing awkwardly in Portuguese one Saturday morning with the lady next door who was a schoolteacher. She asked if Tuca spoke English or if Janice spoke Portuguese to which I answered both questions in the negative. She laughed and said that she often listened to them laughing and talking from over the fence and through the window on Saturday mornings and couldn't understand either of them. I was envious as I had worked hard for ten months by now to learn Portuguese and was still struggling to be understood while my wife was jabbering away with the maid as if they had always known each other.

One evening, we stood in line at an ice cream stand near the plaza. I needed two cones of chocolate and two of coconut ice cream. My pronunciation of their word "coco" for coconut somehow didn't come out right. My order was greeted with hilarious laughter from the crowd in the line behind me as well as my children and the young lady making the cones. I tried a second and third time to get it right with the lady's help but never quite succeeded. Nancy finally explained to me that my pronunciation of the word meant "shit" in the local vernacular. It was humbling to realize that even after trying for months to learn some Portuguese, I couldn't order an ice cream cone without becoming a public spectacle. However, I sought comfort in the fact that most of the people where I worked spoke better English than I did Portuguese. So I didn't get to practice as much as Janice and the children.

We were contacted by LDS missionaries and attended services a couple of times at a small branch where missionaries were depended upon heavily by a few families, none of whom spoke English. Our airfreight arrived from the States with more clothing and household items. We bought several more pieces of furniture, including a TV, kitchen cabinet, floor lamp, and bookcase. Stig came to dinner several times and sometimes visited in the evenings after dinner. We bought a membership in the Clube de Campo (the local country club) where Janice and the children could sun and swim. Besides IAEA employees at CENA, there were visiting scientists and engineers working for Copersucar and for Caterpillar Equipment Company whom we met and interacted with frequently. On Thanksgiving Day, we received a telegram from Jan in Louisiana announcing the arrival of our second grandchild, Dennis Wayne Smith, weighing in at ten pounds and twenty-one inches. Janice again became quite sick at the end of the month, requiring the doctor to come to the house twice. Mrs. Cervalini was a great help in getting us connected with a family doctor. However, we had some difficulty conversing with all the medical doctors, who's English was no better than our Portuguese.

The months of December through March are summer months in southern Brazil, including Sao Paulo State, a thousand miles below the equator. After buying a new Volkswagen Rabbit in late November, our weekends were spent traveling and seeing the country as much as possible. On a January weekend, we spent three days and nights with the Cervalinis at their beach condominium in Guaraja. On other weekends, we drove to see the sights and to shop in Americana, Aguas de Sao Pedro, Barra Bonita, Jundiai, Pocos de Caldas, Rio Claro, Santa Barbra, Sao Paulo, and other places. Americana was founded by disillusioned confederates who migrated to Brazil from the southeastern USA following the Civil War. It was interesting to walk through their cemetery looking for familiar names. At Aguas de Sao Pedro, the children often rode horseback, and there was a waterfall and a favorite restaurant, the Lago Restaurante, on the lake where we dined occasionally and where Nancy and Dan often fed the ducks with bread furnished them for this purpose by the waiters. We all rode the cable car at Pocos de Caldas. There was a beautiful eucalyptus forest at Rio Claro. In Sao Paulo, we shopped at the Hippy Market, dined at nice restaurants, took Nancy and Dan to play in the Cidade do Criancas and to see the Sao Paulo Zoo and Simba Safari. A favorite rest stop between Piracicaba and Sao Paulo was the Frango Restaurant with nearby track for put-put cars that the children frequently drove.

We dined often out in Piracicaba at restaurants including particularly the Brasserie, Falmboyant, Mirante on the riverbank, and the Ortiz. It was in front of the Flamboyant that Nancy fell on the concrete walk breaking off half of an upper middle incisor. She contended with that broken tooth much of the time until we got back to the States. The Ortiz was a downtown eatery for working people and the place where, in my first visit here, I had encountered the sidewalk cartoonist who sketched me wearing a white lab coat with field boots while sitting behind a microscope and contemplating a SCB attacking a cane stalk. The Mirante was on the river bank where we could sit looking out the window at the rapids as water rushed over the rocks in a sloping stretch of the river's bed.

The field surveys were completed in December 1975 after which we were busily summarizing survey data and tending to several field experiments during January. I had managed in less than a year to generate some interest and a lot of opposition to my ideas on integrated pest management (IPM) of the SCB. They just did not want to hear anything about insecticides on sugar cane, although they were being overused there on some other crops. Consequently, I from CENA and Sess Hensley from the USA were invited to speak in Maceio on February 6, 1976, at the Third Brazilian Congress of Entomology on the subject of integrated pest management of the SCB. Our presentations were part of a special session organized by the International Organization for Biological Control, composed largely of entomologists who believed that parasites offered the most practical and reliable means of controlling SCBs and who were generally opposed to any use of insecticides whatsoever. Sess spoke about the IPM program we had developed in Louisiana with its multidimensional approach to pest control through the combined use of resistant plant varieties, conservation of ants, and other beneficial insects and the judicious use of insecticides based on field scouting reports to determine when and where pest populations exceed an established economic threshold.

Much of my presentation was of a preliminary nature as we were still in the process of crunching numbers at CENA and had several pertinent field experiments still in progress. Nevertheless, I said that we had just finished a field survey to obtain much needed and hitherto unavailable information on population dynamics of the SCB in Brazil. Based on our one-year survey, I told them that in spite of all control efforts with parasites, approximately half of their fields in Sao Paulo State had suffered at least 10 percent loss in yield to the SCB, that almost a fourth of their fields had suffered at least 20 percent loss, and that about one-tenth had lost 30 percent or more yield to this pest. I gave them the average numbers of SCB larvae present per hectare of sugar cane for each month of the year and showed how these numbers correlated with periods of maximum sugar cane growth and rainfall. I told them that studies were underway to determine what stages of crop development were most susceptible in Sao Paulo State to SCB damage and to evaluate the possible roles of predators in comparison to parasites as regulators of SCB abundance there. I said that it was our objective to become as familiar as possible with the ecology of the SCB there. We would have more data at a later time to present to these critics but made the most of what we had at the moment. This was my first opportunity to rebut criticism of my ideas with pertinent data collected on site in Brazil. Sess returned with me to Piracicaba and stayed in our home for five days, during which he gave a seminar at CENA similar to his talk at Maceio.

On Sunday, April 4, 1976, I and my family with Tuca left Piracicaba in our Volkswagen Rabbit with our luggage on the top baggage rack. We planned a leisurely tour of southern Brazil to Porto Alegre, a distance of about five hundred miles, pausing to smell the roses along the way. We did not decide on an exact date of return, assuming that we might be gone ten days or more, depending on circumstances, and I did not spend sufficient time or effort planning the details of the trip. Also, cold weather arrived earlier than usual so that there weren't as many roses to smell as we had hoped. This would have been our best chance to see Iguaçu Falls on the Parana River at the Brazil-Paraguay border, but for some reason, we passed up the opportunity.

We stopped the first day at the seaport town of Itajai with its beautiful beaches and palacelike tourist hotels, then vacant except for a skeleton crew of management personnel. Our adjoining rooms were beautifully decorated and furnished, each with a balcony overlooking the beach. Tuca, Nancy, and Dan ran up and down the beach in their bathing suits splashing in the water almost up to their knees. We had dinner brought to our rooms and thoroughly enjoyed this first day out. We drove at least a few hours daily for the next five days, spending the nights successively in Laguna, Porto Alegre, Canela, Lages, and Curitiba before arriving again in Piracicaba on Saturday evening.

At Laguna, we stayed in another hotel on the beach where it was said that whales could surprise you here escaping from the cold waters of the South Pole to this warm-water beach to reproduce and feed their babies. In Porto Alegre, inland from the coast, we had been invited to stay cost free at a camp owned by the relative of an entomologist at CENA. However, when we finally located the place, it was cold and without heat. So we found a hotel for the night. The next night was spent in Canela, noted for its indigenous Indian population. Thursday, we drove on to Lages in the state of Santa Catarina, a historic town named after a Brazilian soldier who came from Portugal in the eighteenth century and became head of one of the royal families of Brazil. We stayed here in a very nice hotel where we also enjoyed a wonderful meal.

Friday, we drove on to Curitiba, the capitol of the state of Parana, said to be one of the best-planned and most progressive cities in the world with a population of almost one million.

Numerous pedestrian malls, lined with fruit trees and flower beds, are the center of city life and are maintained by street children who are paid to do so. We stayed here in the Colonial Hotel and drove the next day to Piracicaba, arriving home at 10:00 PM. This was the only time in her life that Tuca had ever been away from Piracicaba, and she was thrilled beyond belief with the experiences of that week. She was also quite helpful at times with the language, and particularly so on an occasion when I had taken a wrong turn to wind up on a dead-end dirt road. The architecture and people of the extreme southern part of the country were more European, particularly German and Italian, than the vast tropical and subtropical areas to the north in which varying mixtures of these plus Portuguese, Spanish, Dutch, Japanese, Indian, and Negroes predominate—but not to the exclusion of people from every other nation and continent on earth.

Four days after our return from touring the south, I became ill and feverish, and this continued for several days through the Easter weekend, which we spent at the Holiday Inn in Campinas. During May, we all took turns being sick, each for a couple of days at least. Janice was sick a lot during the latter part of May and thereafter, requiring the doctor to make a number of house calls then and also on future occasions. Tuca went to the hospital for surgery and didn't work for a whole week. However, in spite of sickness, we went to Sao Paulo twice during the month for Nancy and Dan to play at the Cidade do Criancas, an elaborate play park for children. When we didn't want Tuca to have to cook and didn't want to dine out either, we took our four aluminum bowls, stacked one above the other in a carrying rack with a lid on top to a favorite "take-out" restaurant for an easy and healthy hot meal. We called this the "dinner bucket" and used it quite often during the cool months from May to September. These months were probably the most socially active period of our lives.

The ladies were constantly visiting back and forth. Janice was learning to play bridge with a group that met regularly. Someone was always giving a party, especially Suzie and Chet Wisner, a plant pathologist retired from the Hawaiian Sugar Planters' Association and now consultant for Copersucar. These friendly seniors occupied a large and stylish apartment in a high-rise building to which they commonly invited a dozen or more guests to dinner parties once or twice monthly.

At one of these parties, we met Deone and Ailene Hewlet, a delightful couple from South Africa and now living in Brazil. They owned a farm, a short drive from Piracicaba, where they grew silkworms commercially by the millions on long benches supplied daily by workers with freshly gathered mulberry leaves. Nancy and Dan were fascinated by the silkworm-rearing processes that we visited on more than one occasion.

One evening in June, I spoke at CENA's postgraduate seminar on the subject of "Integrated Pest Management in Sugar Cane and in General." There were about three dozen graduate students plus several CENA scientists present. Other faculty and students from the agricultural college were notably absent. In my talk, I defined IPM as the intelligent use of two or more control methods combined in an efficient system of pest management to give the most desirable and lasting results in terms of pest control with economy and minimum effects on nontarget species. I explained that it is desirable because it enables us to reap the unique benefits of insecticide use as a last resort without suffering the problems of misuse and to do so with a system tailor-made for each pest problem. I described in detail what we were doing at CENA to develop such a program for SCB control.

I took each of the commonly offered objections to insecticide use and showed that the assumptions on which they were based were invalid. For example, it was assumed that tiny SCB larvae bored into stalks so quickly that there was not sufficient time for contact insecticides to act. I showed them from field survey and insect-rearing data that there was a period of ten days or more for an insecticide to act before the larva bores into the plant. It was commonly said that the SCB breeds continuously throughout the year in Sao Paulo State and that such a long period of dependency on insecticide could not possibly be economically feasible. On the other hand, our field survey had shown that SCB borer larvae were generally abundant for only about six months of the year, and our studies showed that the critical period for protecting the crop from SCBs in any particular field was only about three months. Then I said that the persistence of such myths and misconceptions could be attributed to a lack of critical field observation and experimentation that are and always will be the major sources of information upon which intelligent IPM must firmly rest. I went on to say that a lack of such objective field effort suggests the need to beware of the pitfalls of "whitecoat" and "armchair" entomology.

I then outlined my own vision of what CENA's program in entomology could be: "The rationale being followed at CENA in attacking major pest problems could be a stepwise progression in research, from (1) determining what important environmental factors are already exerting regulatory effects on the pest population, through (2) manipulation of these factors to maximize their effectiveness, to (3) the addition of other factors, which can be profitably integrated into the overall system. When radioisotopes can be effectively employed as a research tool to obtain answers to certain questions, they should be used. However, the total laboratory effort should not be limited to work with isotopes. Instead, efforts should be directed toward synthesis of the complete pest management program. This will commonly require the accomplishment of certain critical studies by CENA personnel, the cooperation of CENA with others to obtain all possible information bearing on the subject, and finally the synthesis of a practical IPM program, which can be recommended to growers and taught to interested individuals through appropriate short courses."

By now, almost everyone realized that the radiation-based sterile male technique for controlling screwworms and medflies would be much more difficult, if not impossible, for SCB control due to the tremendous amount of insect rearing that would be required. Also, problems had arisen in finding a discriminating dose of gamma radiation that would successfully sterilize the male SCB moths without adversely affecting them in other ways. And so in my talk, I continued to hammer away at long-cherished but factually unfounded ideas on SCB biology. I told how we were employing the life-table technique, long used by insurance companies and wildlife biologists, as a means of analyzing data from the field survey and insect-rearing studies in order to divide SCB field populations into age groups or developmental stages to show the amount of mortality that normally occurs during each stage. I said, "Since my life table shows that at least twice as many large SCB larvae die altogether as can be accounted for by parasitism alone, doesn't this suggest that we might need to look more closely at predators, such as ants, which are the most commonly encountered predators found foraging on cane plants in the fields? And if predators are found to be at least as important as parasites, should our efforts in biological control continue to be confined to the rearing and releasing of parasites? I concluded that "naturally occurring predators probably are at least as important as parasites."

I challenged other points of popular belief regarding SCB control, and closed by stressing a point that I had been sent by the IAEA to make, saying that the most desperately needed entomological contribution to the general effort in agricultural pest control is the collection and analysis of field data on insect population dynamics. I said, "Never before have entomologists had such an opportunity to contribute to society and simultaneously to increase the prestige of their profession. However, a special kind of entomologist is needed—one who not only is well-trained in general and applied entomology but who also is well-grounded in all the basic sciences including mathematics. Many people assume that this sort of training should insure that they will always work in air-conditioned offices or laboratories. This should not be so. The entomologists most needed now should be keen observers and tolerant of outdoor work involving sweat and grime. He or she should never be guilty of hiding behind a desk or white coat to avoid fieldwork. The contribution that this entomologist makes will be proportional in large part to the amount of time that he or she personally spends out-of-doors."

As head of the entomology department, Fritz had introduced me to the seminar group in a very brief manner. Now at the conclusion of my talk when I offered to answer questions, I was given little opportunity to do so as he assumed total responsibility for answering questions from students, who spoke in Portuguese and to whom he answered in like manner, monopolizing the time without any effort to translate for my benefit or to allow me to answer. The audience was limited to about three dozen graduate students plus a few CENA scientists and others. I felt that my presentation was timely, appropriate, and also a warm-up for the likely battle to come at the college (ESALQ) later that month.

Following the seminar, I was invited by Sgrillo, another doctoral candidate in entomology, to go for a few drinks with him, Fritz, and Mendonca to the Sarava Restaurant. Artur Mendonca was the Planalsucar scientist who had been the general coordinator for the Third Brazilian Congress of Entomology. Mendonca had visited in Louisiana and was somewhat familiar with the benefits of IPM in Louisiana sugar cane. However, he claimed that Brazilian growers did not have the know-how to handle a sophisticated IPM program, such as ours. I felt that an important role of entomologists was to teach them how.

Two days later, I suggested to Julio that he seemed to have lost some enthusiasm for our work and that I felt someone was erecting roadblocks to reduce the man-hours available for field studies. I also expressed this opinion to the new project manager, Dr. Peter Vose, who passed it on to Dr. Cervalini, director of CENA, who informed the entomology staff that they must give Dr. Long the cooperation that he wants and needs. I appreciated the director's action in my behalf, but there were some things that he couldn't help with, such as Fritz's habitual failure to introduce me to the flow of visitors passing through his laboratory and the fact that, after eight months in Brazil, our wives had never met. My respect for him continued to decrease.

Sunday morning, Janice slept till noon as she had been up past midnight cleaning up from the previous night's dinner party. I took Nancy to the cinema that morning to see American in Paris while Dan played soccer with the neighborhood kids. We drove in the afternoon to Aguas de Sao Pedro, where the children rode horseback for a while before dinner at the Lago Restaurante and the duck-feeding ritual. Prior to bedtime, we had Bible reading and prayers, which Nancy always enthusiastically endorsed. I read from the eleventh chapter of Proverbs, after which it was my turn to lead family prayers. Later in her bed, Nancy called me to their room to say: "Daddy, I love you, and I know that you're trying to be a good daddy to us." I hugged and kissed her and Daniel good night.

Frequent sickness, discomforts of cold weather, and other pressures were beginning to take a toll on Janice who alternated now between periods of cheerfulness and depression. I tried to comfort her but often with little success. We shopped and dined at Electroradiobraz, a favorite department store that we visited often and usually looked first for most of our needs. That evening, the children played with their toys on the floor as Janice did needlepoint in front of the TV and I wrote in my diary.

On some days, I sorted and counted insects from pit-trap collections before coming home at noon to find Janice unhappy and depressed. She often rested in the backyard sunshine and was sometimes cheered by phone calls from a number of friends, including Bernie Robinson and Ailene Hewlet. I often brought the dinner buckets from the Peixinho Restaurante for lunch or dinner. Janice sometimes played bridge with her lady friends in the afternoon, and we often combined dinner at Electroradiobraz with grocery shopping there afterward.

One Thursday was a religious holiday, called Corpus Christi, for all the schools and for CENA. We lazed in bed that morning before getting up late while the children played in the living room. Later, we joined Nick and Alyson Philips at the lakeside in Araras for combined family picnics. Their children and ours rode the paddleboats and played ball, after which we all drove into town to see the street decorations before heading home without waiting for the late afternoon religious procession. At home, we took a walk in the sunset, and Janice made pancakes for supper. The children played with their new toy, the "vai-vem," a toy airplane with two long strings running through it. Dan and Nancy, each holding opposite ends of the strings, forced the airplane to "go and come" between them by alternately and rapidly stretching their ends of the strings far apart. It was a great family day at the end of which the children went to bed on time, Janice worked on her needlepoint, and I reviewed my presentation for the next day's seminar.

Friday was another holiday at CENA and one in which classes would be dismissed early at ESALQ. It was also the day Dr. Neto had scheduled for my presentation before the faculty and student seminar of the entomology department at ESALQ as the last activity of the morning before students would catch buses for home and the weekend. Dr. Gallo, department head, was conspicuous by his absence. The seminar summaries had not been printed for distribution as was normally done. The translation was terrible and was not done by Evoneo as originally planned. When I finished my presentation at 11:50 AM, the chairman, Dr. Nakano, declined to permit questions as he said some students had to catch buses. These all seemed to be efforts to suppress and to play down my presentation in every possible way. Having anticipated this to some extent, I had made it difficult for Nakano to fail to introduce me properly by handing him a sheet in advance with the essential information written for him to read, which he had done with little enthusiasm. I now sensed that I might be fighting a losing battle. As interesting as the work was, I felt that much of my effort would be wasted. And so for the rest of the day, I put all work out of mind and shopped and loafed with my family.

On Saturday, Janice worked most of the day preparing food for a party at Chet and Suzie's that evening. We also contributed a fifth of scotch of which I probably had too many nips, although I could never develop a taste for the stuff. Sunday, we got up late, but happily, for Janice's bacon-and-egg breakfast. Nancy tied balloons on the doorknobs, and they all sang Happy Father's Day as I came down the hall. After breakfast, we drove across the river to admire the street decorations for the Corpus Christi holidays and then to the Clube de Campo where I tried out my Father's Day present of tennis racket and balls. Tuca arrived late for baby-sitting, which caused us to reach the cinema too late for seats. We relieved Tuca, put the children to bed, and spent the rest of the evening at home.

One day, I took off from work a couple of hours to shop for and buy a gold Brazilian good-luck charm for Janice's approaching birthday. That evening, we drove to Tupi, a nearby village, where fire walkers performed their magic after midnight, literally walking barefoot on red hot coals for a distance of approximately thirty to forty feet. We got home the next morning around 2:30 AM. Later that day, I retrieved pit-trap collections from experimental field plots across the river and sorted some of them in the afternoon. In the evening, I shopped for a birthday card for Janice while the children wrapped presents and made cards for their mother. Before I could leave for the field the next morning, Suzie and Arlene came by with more presents for Janice and best wishes for a happy birthday. In the evening, Tuca came to babysit while we dined on lobster thermidor at the Italian La Provence Restaurant in front of a large fireplace with a roaring fire. We held hands and enjoyed a delicious dinner in a cozy and private environment before returning home.

On this Saturday morning, we shopped for clothing with Nancy and Daniel. In the afternoon, Janice prepared food to take to Alyson's apartment for a surprise party for Marge Blockley. In the evening, Nancy stayed home with Tuca while we attended Alyson's party after dropping Daniel at the Didini mansion, on a hilltop behind a guarded gate, to attend Lucianna's birthday party. We danced quite a while at Alyson's place until I wearied and sat down to enjoy another drink. Then Alyson asked me to waltz, which I have never done well and, with one too many drinks, found quite impossible. Alyson said, "I'm sorry, but I don't know which way you're going." I said, "I don't either." And we gave up. Janice was embarrassed and told me later that she had told the ladies that we were ex-Arthur Murray students who loved to dance and that I had really let her down.

Sunday morning, I awoke to a ringing doorbell and a bad headache. Tuca's grandmother had died at 4:00 AM, and she needed to phone relatives in Sao Paulo. Her father came by later to borrow money, which I took to him after changing my last Cr$500 bill. I would have given him more but was obligated to take John Stewart to dinner in the evening. That afternoon, we took Daniel to the preholiday party at the Cigarrinha kindergarten and stayed to watch the children dance. In the evening, we took John to dinner at the Mirante Restaurante on the river. John was a Canadian soil scientist and IAEA expert with the UNDP who was returning home tomorrow. I later took him to see Diva before saying goodbye to him in front of the Esplanada Hotel at about midnight.

On this Monday, my entire day was spent sorting and counting insect pit-trap collections at CENA. Fritz paced nervously around the building most of the day scratching his forearms, red and raw with psoriasis, a sure sign that he was feeling a lot of tension. He scarcely managed to speak until about midafternoon when he stopped at my desk to ask how the bug sorting was coming along. I grunted a short positive answer and continued looking through the microscope. Tuesday evening, we had just finished one of Janice's delicious "fejao" dinners when Peter Vose dropped by for a visit. I believed the project manager was trying to monitor more closely how well Fritz and I were getting along, a situation that had become less well-disguised. I took Peter back to his hotel just before midnight.

For the next couple of days, Julio, I, and the technicians made SCB infestation counts and weighed cane from experimental plots. There was school vacation now for the entire month of July, and Janice was feeling "shut-in" with the weather cold and rainy, and the children home with little to do. She fixed a delicious dinner this evening after which we sat by the electric heater and in front of the TV until bedtime. Saturday morning, we slept late, ate a big breakfast, and left for Campinas about noon to spend a night at the Holiday Inn in a warm room and to sleep late Sunday morning and enjoy a continental breakfast with lots of fresh fruits and yogurt. That day we rode a paddleboat with the children on a lake north of town before returning to Piracicaba in the evening for dinner at the Flamboyant.

On another evening, Tuca stayed with the children for us to attend another dinner party at Suzie and Chet's. Janice truly enjoyed these occasions and made a number of friends among the American, South African, and English group. However, the next day, she was sick again and went to bed with a heating pad for pain in her lower back that had bothered her for some time. Suzie, a retired physician, and nurse Elizabeth Jorgensen thought her trouble might be due to muscle pain caused by cold weather and cold-related poor posture.

The weather was cold and rainy all of this week, including Friday when we hosted a dinner party for invited guests Dr. and Mrs. Cervalini, Dr. Peter Vose, and Diva. Diva sent flowers with a note excusing herself to go to Sao Paulo. The other three guests arrived, Janice's dinner was delicious, the liquor was good, the house was beautiful, but conversation was strained. Janice felt it had been a failure and blamed herself for not being able to "bring these things off like they should be." I told her it had nothing to do with her and that it was all about me, just another evidence of my insecurity in social situations, which in turn prevents others from enjoying themselves. Tuca, bless her heart, swore that she was certain that everyone had enjoyed themselves and the meal without question.

With the weather continuing cold and rainy, we slept late Saturday, and Janice went back to bed after breakfast with continuing back pain. The children and I went to buy a kite and fly it at the college campus. By evening, Janice's pains had increased to include headache and fever. Sunday morning, I called Dr. Marcelo Tricca. His car had been wrecked and was in the shop. So I drove to his home to get him and brought him to see Janice. He prescribed an antibiotic plus a muscle relaxer and a pain pill. He also wanted urine and blood tests to be done the next day. By evening, she was feeling better but was still in pain when Dr. Tricca returned to see her again. He told me how much he appreciated the scotch whiskey, brandy, and liquor I had given him earlier and that he had left some friends visiting at his house to come and check on Janice. He had done a great deal for us since our arrival last October, making five or six house calls already and seeing me for an office visit without ever permitting us to pay anything at all for his services. I could not understand it but had been pleased that morning when he accepted five bottles of imported goodies from Copenhagen, which would have cost him about Cr$1500 or one hundred fifty dollars. Marcelo Tricca regularly read medical journals in English but seldom had opportunities to speak the language, which may explain his willingness or desire to spend as much time with us as he did. Regardless of his reasons, we felt blessed by this relationship.

As the new workweek started, Tuca was nursing the sick at her house while I fixed breakfast at ours after going to the medical laboratory to fetch a bottle for Janice's urine sample and to pay for lab work. A nurse came by during breakfast to collect a blood sample from Janice but forgot to take the urine sample with her when she left. So I passed by the lab myself and left it there on the way to Dona Karen's house where I left Nancy for her Portuguese lesson before returning home for a shower and drive to CENA. I returned home after sorting insect collections for about three hours to find that Elizabeth Jorgensen had come and taken Nancy and Dan to fly kites with her children. And now some other friends of Janice's, Allison, Mina, and Gwen arrived to visit at the same time that Dr. Tricca called to say that the urine was normal and other results would be available tomorrow. That evening, I brought dinner home in buckets from the Pexinho Restaurante. Janice's back pain continued the next day, but without fever. We both worried about what her trouble might be and hoped she would not need any surgery before getting back to the States.

I fixed breakfast, took Nancy to Dona Karen's, shopped for groceries, took Dan to fly his kite, and checked mail at CENA. After lunch, Peter Vose and I went shopping in preparation for his wife's arrival from England. I finished sorting insect collections later in the afternoon and returned home just before the arrival of Elizabeth and Cida to visit Janice. A few minutes later, Dr. Tricca came again, took her off the muscle relaxer, and prescribed a pill for her cough. He was still uncertain about the cause of her problem. The children and I went to Electroradiobraz for dinner, to fill the prescription for their mother, and for chocolate ice cream and cake at her request. She ate the dessert slowly, commenting that it was not as good as the last she had enjoyed, and sat with us for the rest of the evening. This was the first time Janice had been out of bed in three days, except to go to the toilet.

The next day, she was back in bed all day, during which I divided my time between CENA and home. It was good to have Tuca back at work again. Dr. Tricca came in the evening as usual and told me that he was beginning to suspect rheumatic disease as the possible cause of Janice's back pain. Such a diagnosis begged for something more definitive, but that was as close to a positive diagnosis as he ever gave us. In retrospect, that may have been the best anyone could have done. She had suffered from a streptococcal infection and was showing a lot of emotional instability, which sometimes accompanies rheumatic fever.

The next day, we went to the hospital for x-rays, which showed some liquid in her lungs, causing the doctor to add flu and reactions to medication as additional possible causes of her problems. Dr. Tricca now wanted us to get an expert opinion from Dr. Piazzo. Upon returning home, a neighbor lady came to say that Daniel was digging in her yard and had broken a waterline to her house for which I should reimburse her for repairs. Susie, Allison, and Arlene came to visit and brought Janice souvenirs of southern Brazil. The following day, I paid the neighbor Cr$40 for her plumbing repairs and told her I was sorry it had happened, that in the future she should send Daniel home whenever he came to her house, and I promised to do the same when her children came to mine.

Janice was in a lot of pain as I accompanied her to see Dr. Piazzo, where we spent an hour and Cr$300 telling him about her family, her past medical history and present symptoms, but getting no opinion, examination, or recommendation. The good doctor wanted to see her again on Monday. When Monday arrived, his secretary called to postpone the appointment until Tuesday at 2:00 PM. At 12:30 PM on Tuesday, his secretary called again to postpone until the next day at the same time. Janice was almost ready to go when the call came and had no desire for another appointment. We both agreed. I told the lady to forget it and that we needed no more appointments with Dr. Piazzo.

Saturday, July 17, Janice felt much better. Tuca believed that the sickness was caused by a curse pronounced by a spiritualist, probably being paid by someone I had offended at CENA or at ESALQ. She had heard from a friend that both Dr. Tricca and Dr. Piazzo are also spiritualists, and that, recognizing Janice's problem now, they needed no further examinations but only needed to pronounce a counteracting blessing, which they now had obviously done since Janice was so much better today. I took the family to Aguas de Sao Pedro in the afternoon where the children rode horses before dinner at the Lago Restaurante. We slept late Sunday and went for a ride that afternoon to fly kites, pick up stones, and take home dinner in the buckets. The past week had not been a good one. While Janice seemed better physically, she was suffering emotionally. She thought at times that I liked to see her suffer and believed that we were in a trap that prevented either of us from doing the things we really enjoyed. However, at other times, she also admitted that she didn't know what I could do to make things better.

Monday, July 19, Janice felt better than she had for a long time. Dr. Tricca came in the evening and was happy to see her almost pain free. Tuca had almost convinced her that all her recent health problems stemmed from a voodoo-type curse. Life suddenly got better as the curse was lifted. We had dinner at home five nights this week and dined out at the Ortiz one evening. Saturday, while the girls and I were shopping, Daniel disobeyed Tuca by riding with a neighbor to his farm in the country. Upon his return, I lectured and spanked him and supervised his shower before walking with him to the Clube where we batted tennis balls on the practice court. That evening, we fried fish for dinner and then let Nancy and Daniel go to Toto's birthday party next door while we went to another party at Suzie and Chet's. We left our party before midnight to retrieve the children from Toto's house where we went in to visit and have refreshments before going home at midnight.

I spent the entire week at CENA doing statistical analyses of experimental data in preparation for publication. Although a written report summarizing my work with conclusions and recommendations was required, publication was not a necessity. When such publication was done, it was commonly done in scientific journals. In this case, I decided to publish in a technical journal, *Sugar y Azucar*, widely read by sugar industry people, particularly in the Western Hemisphere. I had hoped that our recent work might have some impact on agricultural thinking and practice in the near future and entitled the paper "Basis for Practical Use of Insecticide in Management of Populations of the Sugarcane Borer in Sao Paulo, Brazil." However, even my friend, Chet, suggested that my significant yield increases might be due to plant stimulation by the insecticide rather than insect control. Although such effects have been reported for some other chemicals by workers, including me, there was no reason to believe such in this case. I attributed Chet's remark to his background in plant pathology. After all, SCBs don't cause disease. They just devour the crop.

Sunday night, we had dinner with Suzie and Chet and came home nauseated. Janice vomited during the night and felt bad for several days afterward with dizziness and pain in her neck. Tuca announced on Monday that she would have to quit work due to problems at home and then said the next day that she would continue with us somehow until October 2 when Janice and the children would leave for Louisiana. It was already August, and we had not yet been to Rio. The first draft of my paper for *Sugar y Azucar* was completed. Nancy had been under the weather for a few days, but Dr. Tricca thought it would be safe for us to leave the children with friends (Nancy with Dona Karen, and Daniel with Tuca) while we went to Rio for a weekend. School had started again after the July vacation, and the children were excited and happy about that.

We departed Friday morning for Rio de Janeiro in our white Volkswagen Rabbit to be by ourselves for the first time since arriving in Brazil. Janice's eyes filled at the mention of their approaching departure in October, but we had both agreed that it would be the best all-around solution to several problems, including school for the children that would start in September back in Louisiana. I would join them there in November after completing my contract. We arrived in Rio Saturday afternoon and checked into the Rio Sheraton before dining on prawns at Maxime's restaurant on Ipanema beach. Sunday morning, we went to the Hippy Market and in the afternoon visited the Pao de Azucar and the Corcovado (Christo). That evening, we dined in the Sheraton's Sarau restaurant on baked whole red snapper. Monday, we toured H. Stern's where we heard explanations and watched gem making in progress. I bought Janice a topaz ring and gold charm at Stern's, plus two citrine topazes, a smoky topaz, and an amethyst at nearby souvenir shops. We shopped on Copacabana Avenue after lunch and enjoyed a late dinner and floor show at Las Brasas restaurant. We had breakfast in the room each morning. She slept like a rock to the sound of pounding surf just beyond our open doors and balcony facing the sea. I lost a couple of hours getting on the right road home by taking a wrong turn and having to ask for directions. When we arrived home Tuesday at 9:00 PM, Tuca had dinner ready and waiting, had washed the living room drapes, cleaned the mildew off the bedroom walls, and had the house sparkling clean. She had had no trouble with Daniel, and Nancy had enjoyed the weekend with Dona Karen. It was a leisurely, restful, scenic, entertaining, and romantic weekend, during which most of Janice's symptoms of neck pain and other discomforts disappeared.

By mid-August, Nancy was still ailing with a chronic cough, weight loss, and lack of energy. Dr. Tricca had been treating her for a common cold, but she continued to fade and lose weight in a way that was frightening. While talking with him about the situation at his office one day, it occurred to us that rapid weight loss with diarrhea might be due to her inability to absorb food efficiently. Within minutes, we had a wet mount from a fecal sample on the microscope. Dr. Tricca looked at it, smiled, and told me to take a look. There were hundreds of little protozoan cells milling about apparently looking back at me like tiny Halloween masks. She was infected with *Giardia*, commonly transmitted by fecal contamination of water supplies. She improved spectacularly after only a few hours following medication and was her old self again after several days.

During the last week of August, we had a dinner party for Chet and Suzie and the Jorgensen family. On another day, we took Suzie with us to Sao Paulo to the Praca da Republica and to shop at the Hippy Market. The final corrections were made on my paper; it was copied for local distribution and sent to the editor of *Sugar y Azucar*. Airline reservations for Janice and the children to return to the States October 2 were confirmed. As the time approached, Janice felt increasingly reluctant to leave. She had made many friends among the English-speaking "extranjeiros" here and was continuously busy going to dinner with friends, having friends over for dinner, playing cards, etc. I had even developed a temporary interest in gemstones and minerals, probably out of a desire to take some of Brazil back home with me. I read two books, "Gemstones" by E. H. Rutland and "Discovering Rocks and Minerals" by R. A. Gallant and C. J. Schuberth before contracting with a professor of mineralogy at ESALQ to prepare a labeled collection of one hundred sixty Brazilian minerals that I would take back to the States.

On a Friday in early September, I drove the family to Cabo Frio where we spent two nights in the Malibu Palace Hotel on the beach under cloudy skies and rainy weather. We drove on to Rio on Sunday for two nights in the Rio Sheraton. This trip was a combination final sight-seeing tour for the family and celebration of our twenty-third wedding anniversary (September 11). Monday, Janice and I with Nancy and Dan rode the cable car up to the Pao de Azucar and visited the Christo, where we bought souvenir dishes embossed with our pictures. The sun finally came out Tuesday morning to permit sunning on the beach and playing around the pool. We left after lunch for the return trip home, arriving the next day with substantial additions to our souvenir collection. Upon arrival, Janice called her parents in North Carolina where her mother and aunt Chloe answered the phone. She started telling them excitedly about her plans for returning to Louisiana by way of North Carolina. Then considering that school had already started, for which the children needed to get home soon, she decided to go directly home to Louisiana.

One evening in late September, we attended a going-away party for the Longs at Allison's apartment next door to Suzie and Chet. It was a gala event and well-attended by our American and South African friends. Janice was presented a beautiful brass candelabra set that now decorates the sideboard in our dining room.

Mission Completed. Saturday, October 2, almost exactly one year after arriving as a family, we left Piracicaba for Sao Paulo and the international airport in a CENA vehicle driven by the friendly motorista, Jabao, with Daniel and Nancy beside him in the passenger seat and Janice and me in the rear. The children talked excitedly and continuously with Jabao during the trip and in the terminal until they and Janice boarded their plane. We parted at 8:10 PM with hugs and tears and with Nancy talking so rapidly and expressively with Jabao that she sounded to us like a native Brazilian. They arrived safely the next morning in Louisiana and in Thibodaux about noon. I called Sunday evening to verify that all was well. The children were excited about going home, but Nancy had also shed some tears over leaving Brazil. I had asked Diva earlier to look into the possibility of my detouring by way of Lima, Peru, and the Galapagos Islands on my return home next month. Later I realized it was almost 10:00 PM, and I was still in the laboratory at CENA writing in my diary. I hated to go back to that empty house at night.

Chet and Suzie Wisner insisted that I come to breakfast with them Sunday morning, and that evening, they and I had dinner with John and Arlene. The next morning, the Wisners left on a South American tour, and I began to take meals regularly at their apartment for the next two weeks while they toured. They had two full-time employees for cooking and housekeeping and insisted that my daily presence there would be beneficial to them as well as me.

The next three days were spent in the office writing my final IAEA report to the government of Brazil. On Friday, I gave it to Peter Vose and Diva for review and typing. Peter said that it was well-written and absolutely true but was so critical of the local establishment that it would be classified as "restricted" if left as it was. I told him that I would welcome suggestions for making it more diplomatic but that the hard facts should be made clear.

The final summary report of fifteen pages was entitled "Research Programme on Ecology and Management of the Sugarcane Borer, *Diatraea saccharalis*, in Brasil" and contained all the recommendations I believed were important. In addition to this, the full report contained copies of all publications that resulted from work done during my mission. I made the following general recommendations to the government of Brazil for the future handling of entomological problems:

> There is need for a better balanced program in sugar cane pest management research, which should include simultaneous efforts in insecticidal control, biological control, cultural control, and plant resistance to insect pests (varietal control). Far too much of the total effort in sugar cane entomological research in Brazil appears to be spent on biological control by parasite releases. There is too little foundation of basic information to support these biological control efforts. More laboratories are built each year in order to release more parasites, and yet no one really knows how many parasites per hectare to release, when they should be released for maximum effectiveness, or even with how much confidence any effect can be predicted. Unfortunately, the sugar cane entomologists of the country, upon whose recommendations both government and often industry depend for decision making in this field, see little wrong with this approach.
>
> More entomological research in other areas of agriculture is also sorely needed. For example, estimates from informed sources suggest that much more insecticide than necessary is used on the average to control soybean insects in southern

Brazil. If this situation is permitted to continue in soybeans and possibly other crops as well, there probably will be serious environmental effects.

A concerted effort should be made to recruit applied entomologists for research and extension work with strong interests in field entomology and pest management, and with sound training or experience abroad in these fields. People with this background and interest generally should be the leaders of entomology research. In too many cases insecticides are grossly misused or overused, or they are not used at all where some use is needed. Correction of this situation will require more entomologists and more entomological research in pest management. There are today less than fifty active research entomologists in Brazil, relatively few of whom are doing field research, and perhaps no more than three hundred professional entomologists in the entire country. By comparing the populations of Brazil and the United States and the respective memberships of their leading entomological societies, it is estimated that the United States has approximately fifteen times more entomologists per capita than Brazil. Furthermore, in Brazil, there are more than twice as many soil scientists, four times as many geneticists and plant breeders, and approximately twice as many plant pathologists as entomologists. However, a survey of several 'chefes das casas de agricultura' (extension specialists) indicates that approximately 30 percent of all requests received by them for assistance or information are of an entomological nature.

> Theoretically, it should not be surprising to find the sciences of crop production developing slightly ahead of those of crop protection. However, if the latter do not catch up, much of the former becomes wasted effort.

In addition to these general recommendations, I made specific recommendations relative to the government's work program in entomology and to the specific needs for further work on pest population dynamics, insecticide field trials, identification and augmentation of beneficial species, continued biological control studies with parasites, and to the needs at CENA for supporting such efforts.

During early October, I tried repeatedly to interest the CENA entomologists in visiting with county agents to obtain general information on practices and problems in other crops and to locate cooperating sugar mills for continuing our field research program. These suggestions were received with lukewarm approval but little follow-up action. Fritz said that his wife knew an agricultural economist in Sao Paulo who could give us all of this information. There simply was not much interest or desire to get involved directly with farming people in on-the-farm solutions to farming problems.

On Friday, October 8, I moved out of what had been our home for a year at 1123 Rua Edu Chaves as another family moved in. I paid Tuca Cr$1,000 for helping me to pack and load our belongings for shipment to the States, hugged, and kissed her goodbye and promised to pass by her house again before leaving. Saturday evening, Peter and Jean Vose had me over for dinner, and the next evening, it was dinner with John and Arlene Manchini. Most of Sunday was spent at CENA writing in my diary and writing letters to family and especially to Jan. I had wished so many times that she could have been with us during the last twelve months.

During the next week, I finished making corrections on two manuscripts for publication in the journal, *Ecossistema*, one on the effects of SCB damage on the weight and quality of sugar cane and the other on the effectiveness of two insecticides against a sugar cane root spittlebug. I also visited Dr. Salles at the Casa da Agricultura in Piracicaba and Dr. Peixoto of the Bom Jesus sugar factory with regard to finding cooperators for continuing the field studies. People in the industry, who actually worked at producing cane and sugar, usually were interested in participating in related research when contacted. I finished and mailed a newsy long letter to Janice telling her the name of the ship on which our freight sailed from Santos on October 14.

On Saturday, I took the Jorgensen family to Sao Paulo where we visited a bookshop and stone shops before noon and where I bought a nine-carat rubylite birthstone for Jan. In the afternoon, we visited the Sao Paulo Museum of Art before the Jorgensens were deposited at the home of their Scandinavian friends, and I went to spend the night at a hotel near the Praca da Republica. Sunday morning found me at the Hippy Fair on the Praca at 9:00 AM where I finished Christmas shopping and bought a perfect trilobite fossil for Cr$1,000. I have never really wanted to know how much I may have been screwed on that deal. The Jorgensens and their friends lunched with me near the Praca, and we drove through a rainstorm on the way home to Piracicaba.

Julio and I visited Dr. Jose Roberto, at one of the nearby sugar factories, who showed interest in cooperating on further field studies with insecticides. I went fossil hunting with Roberto Forti and Sorem Jorgensen to rock quarries near Rio Claro where we found a couple of hundred-million-year-old fossil remains of the Permian lizardlike reptile, *Mesosauro brasiliense*, in two different quarries. Forti found a third specimen so complete and perfect that I bought it from him. I called Janice to tell her to meet me at the New Orleans airport November 8. She was happy to hear the news and said that I should be getting a twenty-page letter from her soon.

During late October, I enjoyed a flurry of dinner invitations from friends, all wishing me goodbye and good luck. I took a copy of my final IAEA report to the government of Brazil with some other books for binding before selling my car for Cr$36,000. A jeweler mounted Jan's rubylite birthstone on an 18K gold band, and Janice's letter arrived describing various problems, including some with the business. One day, I sorted and packed books, papers, and correspondence at the office; and that evening, at Nick and Allyson's apartment, watched Dr. Gallo on TV tell of the virtues of biological control of the SCB by parasite releases.

A letter came from Sess Hensley telling of his decision to return to the USDA sugar cane research laboratory in Houma now that his old boss had retired and of his desire to see me back at LSU if I so wished. Actually, the idea was not without appeal. I was getting physically and mentally tired of commercial enterprises and would have welcomed going back to nearly full-time research. Also, the heavy teaching loads in freshman zoology at Nicholls State were more chore than enjoyment. However, realities would not favor such a move. LSU would have had to pay me as much as any other of their senior faculty, which would have amounted to reduced income for us without my consulting business. Also, Gene Reagan had earned his MS in entomology under Sess at LSU and was now about to finish his PhD at North Carolina State with the hope of returning to LSU, where he could be hired at a beginning faculty salary. He and Sess could cooperate in research benefiting the sugar industry while my consulting business, which had pioneered the IPM practice of properly scouting sugar cane for SCB control, would continue to serve the industry well by staying put in Thibodaux. And this is the way it all worked out as Dr. Reagan assumed responsibility for entomological research on sugar cane at LSU the following year (1977).

Peter and Jean Vose invited me, the Jorgensens, Diva, and Cida to dinner, a kind of official farewell for me. Because of concerns for time and money, I had given up returning home by way of the Galapagos Islands and elected instead to stop off in the jungle city of Manaus between Brasilia and Mexico City. On Friday, I shipped home my books and papers via airfreight, picked up airline tickets and Jan's ring from the jeweler, bought beer for the entomology laboratory personnel, changed my money to U.S. dollars, and dined alone in my apartment. Saturday and Sunday, I had drinks, pizza, refreshments, coffee, and dessert with three different sets of friends and acquaintances.

Monday, November 1, I entertained Chet and Suzie Wisner along with Peter and Jean Vose with dinner at the Sarava restaturant. I had dinner the next evening at the Wisner's apartment, spent the night there, and had an early Wednesday breakfast at 5:00 AM with them before the CENA motorista came at 5:30 AM to take me to the airport. Jabao had been given instructions to stop by Diva's house before leaving for Sao Paulo. She accompanied me on this final trip to Sao Paulo where we parted for the last time with a hug and a few tears. I arrived at the UNDP office in Brasilia before lunch, gave them my report, and finished debriefing in the afternoon. The next morning, I checked out of the National Hotel and took a noon flight to Manaus, the jungle city of over a million people near the union of the Rio Negro with the Amazon River, a thousand miles up river from the Atlantic Ocean and almost exactly on the equator, a city literally rising out of the middle of the world's greatest rain forest.

On arrival, I checked into the Hotel Amazonas in time to take an afternoon walk that took me to a souvenir shop called the Humming Bird, filled with Indian blowguns, spears, ceremonial staffs, and other curiosities that they sold and shipped anywhere in the world for a single price. I exchanged $150 for cruzeiros with some of which I bought souvenirs and left instructions for shipping them to Thibodaux. A walking tour in the afternoon took me past the world-famous Teatro Amazonas, the Manaus opera house built by wealthy rubber barons over a hundred years earlier. It was built piece by piece in Europe and shipped to Brazil for assembly. I walked in the public market with a red roof, designed by Eiffel and also built in Europe and shipped to Manaus. I caught the 5:00 PM boat for January Island and a night in the floating lodge on the Amazon and was surprised to find that I was the only overnight guest at the lodge.

Saturday, I bought a tour package that included a walk in the rain forest jungle, a visit to an Indian village to see ceremonial dances, and a boat ride to the "meeting of the waters" of the black Rio Negro with the muddy yellow Amazon. We returned to Manaus where I checked into the Tropical Hotel for dinner and sleep. Sunday morning, I walked around to take a few pictures before checking out of the hotel and taking a taxi to the airport. Varig flight 872 was scheduled to depart at 1:50 PM for Mexico City and finally took off at 3:35 PM. Although I had visited Receife and other places in northern Brazil, this trip to Manaus was quite interesting and satisfied my curiosity to see a rain forest at the equator. I checked into the Hotel Sheraton in Mexico City late Sunday night.

14

LATE SEVENTIES AND EARLY EIGHTIES

Monday, November 8, 1976, I arrived in New Orleans at 4:40 PM aboard Eastern flight 906 for a happy reunion with wife and children whom I had not seen for a month and four days. The next day at home, Janice helped me to comprehend the things that had happened during our absence—the failure of young couples with young children to keep up with the care of a big yard and house was the least of our problems. When I walked into our office on Himalaya Street, there was dust and empty containers stacked everywhere. Copies of invoices showed that Dennis had kept up pretty well with treating homes and businesses for control of crawling insects. But Dennis was an engineering student at Nicholls State and had no real interest or enthusiasm for spraying crawling insects. Some callbacks had not been answered for more than a month.

Charles Taylor had done the summer field scouting for farmer customers without losing any of them. However, Charlie had paid himself full salary the whole time without doing much of anything for the rest of the year but drinking and running around on his wife. She had filed for divorce, and his drinking had been out of control for a while. I fired him immediately, took over the company checkbook, and notified the bank to no longer honor his signature. Charlie had come from a good family, had been a good student, a quick learner, and a hard worker; and I had given him stock in the company to encourage him to stay with me permanently. Now he wanted payment for his stock for which I shelled out two thousand dollars more to get rid of him. Ironically, at about this time, the UNDP wrote to enquire if I might be interested in a one-year assignment in Bangladesh to which I responded negatively without hesitation.

I now realized that without an additional management person in the pest control portion of the business, I was wasting time and money trying to do too much with too little. Field scouting and consulting had grown to require a field assistant, and my teaching salary at the university was still an important part of our income. Also, I had to face the fact that the pest control business was suffering for lack of salesmanship and someone with more business and people skills than I had and that being the most knowledgeable and best qualified to do a job does not necessarily insure that one will get to do it. We were doing a termite job once or twice a week and spraying lawns, shrubs, and trees during part of the year. But after five years, income from this work was not covering the costs of doing business. Agricultural consulting was paying to keep the pest control service afloat. And so my business, which had begun strictly as agricultural consulting in 1965 and had added pest control services in 1972 as Long Pest Control Inc., now changed in name to Long Pest Management Inc. in 1977 and again was confined to agricultural consulting. I sold the pest control part of the business with the equipment to Charles Beasley, owner of Beasley Pest Control Inc. in Houma, Louisiana, and was as happy and relieved to get rid of it as Charles was to buy it.

By now, Jan and Dennis were divorced; and Jan was living with us again in the big house with her three children—Holly, Denny, and Heather. Heather was nearly a year old when they moved back to be with us. Jan attended classes at Nicholls State with a part-time job at the TG&Y. Holly and Denny attended nursery school for a year and then first grade at Saint Joseph Elementary. Fortunately, we had not forgotten how to change diapers, and I usually took my turn. Heather sometimes rode with me to my office at Nicholls. I still chuckle remembering how her curiosity prompted her one day to pull the fire alarm in Gouaux Hall causing the classrooms to empty while fire inspectors searched in vain for a problem. At another time, the upstairs bedroom, which Dan shared with Denny, began to smell strangely like urine. Further investigation revealed that someone had been peeing in a leather souvenir trash can from Brazil, soaking it and the carpet beneath, for fear of walking in the dark hall to the toilet. It wasn't funny at the time, but I've laughed about it many times since, remembering how I used to empty my bladder through a window screen in the second-floor bedroom at the farm to avoid using a nearly filled pot in the room and having to go outside at night.

Now that I didn't have to worry anymore about the business of termite, cockroach, and horticultural pest control, I could better focus my attention again on crop consulting. I was now a little annoyed by the fact that LSU had slightly altered their SCB control recommendations that I had written in 1964. Those earlier recommendations had stated that fields should be checked weekly from mid-June through August, and my published research had shown no benefit from insecticide treatments after August. Then in 1972, without further research on the subject, the LSU recommendations were altered to suggest the need for weekly field scouting through mid-September. And so in 1977, I conducted another replicated field experiment that again showed no benefit from mid-September insecticide treatment and that supported my general belief that protection of the crop from SCBs is not helpful after the cane has reached at least 70 percent of its final height. This was later published (*Sug. Bul.* 60(21):9-11) to help alleviate concerns about late season SCB infestations.

In the late seventies, Shell Chemical Company obtained a label for the use of Azodrin insecticide on sugar cane. It was cheaper than Guthion, which was most commonly used at that time and killed some insects more rapidly and completely due to its systemic effects, allowing it to kill SCBs both before and after they entered the cane stalks. I had used Azodrin in test plots and knew that it performed better than Guthion in many situations and would cause no fish kills for which there was concern following Guthion use. However, because of greater hazard to birds roosting in cane fields and to other concerns, the LSU extension and research entomologists advised against its use. Since I didn't agree with them, I began recommending it in 1979 to clients who used it with great success. Other consultants then began to recommend it, and the LSU entomologists started a campaign to halt its use because of occasional bird kills, primarily of red-winged blackbirds that many farmers considered pests of their fruit and grain crops. I believed that the LSU workers also were annoyed by the fact that a consultant would dare to suggest anything not in line with their recommendations as evidenced by the fact that they sometimes now spoke publicly of consultants in a derogatory manner.

Dr. Denver Loupe, LSU extension sugar cane specialist, told consultants at their pest management workshop in Monroe, Louisiana, February 25, 1980, that "consultants do not make management decisions for farmers" and that they should "stick to counting bugs." Only one with the arrogance of Denver Loupe would have made such a statement, and of course, it didn't go unchallenged. A consultant, Dr. Dick Jensen, rose to say that the time would come when consultants would contribute input to the formulation of state pest control recommendations. I supported Dick's statement with the fact that the LSU cooperative extension service could not afford to ignore the wealth of experience and firsthand information available to it and to the citizens of the state from consultants.

I had been asked by Dr. Calvin Viator, then secretary of the Louisiana Agriculture Consultants Association (LACA), to represent consultants at this pest management workshop as a member of a panel to discuss sugar cane insecticides. Other members of the panel were Dr. Sess Hensley, by this time research leader for crop protection at the USDA sugar cane research station in Houma, Louisiana, and Dr. Gene Reagan, assistant professor of entomology, in charge of sugar cane insect research at LSU. This is the only instance I can remember in which Sess and I went head to head against each other in open public debate. Each panel member spoke for about fifteen minutes after which there were opportunities for rebuttal. Hensley and Reagan emphasized the importance of growers following recommendations based upon LSU and USDA research and the potential pitfalls of not doing so.

After some preliminary remarks about the inevitability of differences of opinion among intelligent and independent thinkers, I said, "Since I recommended and was responsible for more farmer use of Azodrin last year than any other sugar cane consultant in Louisiana, it is perhaps reasonable that I should represent these renegade consultants on this panel." I pointed out that good things, like the U.S. Constitution, had been wrought in the fires of conflict where the participants adhered to rules enabling them to avoid mutual destruction. I recalled that the Louisiana Pesticide Control Act of 1975 no longer required consultants to follow extension service recommendations as did the old Louisiana Horticulture Law and Regulations of 1965. I suggested that it would help avoid undesirable consequences of conflict like this if we would all accept and follow a code of ethics and said:

> "It is obvious that we do not all agree about the need for ethics. For example, an unsolicited research professional wrote a letter advising a client of mine that he was being ill-advised by me to use Azodrin. Nevertheless, the client followed my recommendation. I could cite other similar examples. In my opinion, this researcher violated some basic professional ethics. I have learned through more than one source that inaccurate and biased statements have been made by more than one institutional employee at farmers' meetings for the purpose of steering farmers away from Azodrin use on sugar cane. Farmers at a recent meeting were told by a research entomologist that the use of Azodrin in an area last summer had resulted in more insecticide application than should have been required. This is absolutely untrue. Let's stick to the facts."

I then enumerated the facts in terms of arguments for and against Azodrin use, citing pertinent research and experience, and weighing the pros and cons before stating that I would continue to recommend and my clients would continue to use both Azodrin and Guthion. I expressed respect for the institutions and personnel of LSU and the USDA and a dependence upon and need for them.

> "However, I will not be intimidated by any individual or institution. If those of us with common interests cannot work together and disagree together in a spirit of mutual respect, then we will work separately and in opposition. Heaven forbid! If not, we'll all be losers. But consultants will not be the biggest losers. There is no way that a consultant,

who has served his customers well for years, can lose a significant amount of business because of an honest difference of opinion with a research scientist or extension worker."

In closing, I made a plea "that we hereby resolve to cooperate with one another and that when differences of opinion arise between us, we sit down and discuss them and seek to reconcile them in a spirit of mutual respect." In spite of this plea, the rebuttals became heated and the entire panel presentation lasted at least an hour and a half. When it was finished, I felt good about it all and was congratulated by a number of people at the meeting as well as by letters and thanks afterward.

In spite of its economy, effectiveness, and record of safe use on sugar cane, the use of Azodrin insecticide was eventually prohibited on sugar cane and most other crops during the middle eighties by the Environmental Protection Agency (EPA). Environmental groups, like Friends of the Earth, had begun to target the agricultural use of pesticides in earnest with the intent of outlawing the use of all chemical pesticides in agriculture without regard to toxicity, mode of action, economic impact, or other reasonable considerations supporting their use. In 1980, I wrote letters to the EPA, the Federal Aviation Administration (FAA), and to the congressmen and senators handling their budgets, urging them to use common sense and fairness in responding to petitions from Friends of the Earth and supplying them with considerations and facts in support of agricultural and economic concerns.

During the first twelve years of my work as an agricultural consultant (1965-77), I had steadfastly resisted offering anything more than entomological services in the belief that I had no business posing as an authority in areas outside my specialty field of primary training that was entomology. Although I had a minor in soil science, some credit hours in plant pathology, and I studied all the sugar cane crop management recommendations annually, I didn't feel as qualified to advise clients on non-insect matters. Also, early in my career at LSU, I had been outraged by criticism from scientists outside my field and whom I had subsequently given a lesson in minding their own business and staying out of matters about which they had little or no training or experience. However, now as a consultant, I was being asked questions constantly about weed control, sugar cane varieties, soils, and fertilizers to which I had always deferred to a real authority or promised to find and bring them an answer.

My main competition in business now was from a native south Louisiana son, Dr. Calvin Viator, who was raised on a sugar cane farm, earned a PhD in plant pathology, and joined the biology faculty at Nicholls in the late 1960s. Calvin had worked for years as a sugar cane field scout for Al Dugas before finally going on his own in 1970. His business thrived for a variety of reasons, one of which was that he was a generalist and had always offered his clients answers to most of their questions. By the late seventies, I finally began to realize that I also would have to become more of a generalist to be more competitive. After all, who would want a family doctor who only treated colds and flu? And so in 1977, I took all the state exams required to be certified in all additional categories in which I might offer services, including soil science, plant pathology, weed control, and others. I began to offer soil testing with fertilizer recommendations and to make recommendations on weed and disease control and variety selection.

Calvin and I also differed in other respects. He was a good promoter with ambitions to rival his old boss, Al Dugas, in growing a business as large as possible. I had no desire to be responsible for more farmers' fields than I could personally see if necessary on a weekly basis. I was not a great promoter, and I was not comfortable delegating many consulting decisions to others. Early in my consulting career, I had declined to accept help from several of my customers who each had offered to help solicit more business for me in their neighborhoods. I was afraid at that time of getting too big to provide proper service. Also, my approach to borer scouting of fields had begun with a philosophy of intense micromanagement that was very time-consuming and would later prove to be sometimes impractical while Calvin's approach, learned under his old boss, was to paint field maps for insecticide treatment with a much wider brush. For a while, I saved growers more insecticide dollars although they often had to call aerial applicators more often to treat smaller acreages. As time went on, it seems that Calvin tried harder to do more intensive field scouting while I learned to paint with a little wider brush until by the eighties, our services were more similar in this respect.

My natural tendency was to focus more on the sciences of entomology and crop management and less on the business. My business grew mainly through references from satisfied customers. However, by the late seventies, I was routinely sending letters to all growers on the mailing lists of several sugar mills but not doing much follow-up visitation. On the other hand, Calvin was competent with the scientific aspects of consulting but he also worked at promoting his business by active face-to-face (door-to-door) selling; and by alliances with banks that handled farm loans, and with attorneys who catered to lawsuits involving disputes between growers or landowners and oil or pipeline companies; and through other business, social, and family relationships. As a professor, he seemed to actively court the students of potential customers.

His aggressive style may also be partially illustrated by an event at a meeting of the Bayou Young Farmers Association in Napoleonville, Louisiana, on a Monday evening in April of 1983. He and I both were present, and each were given time to describe our services and discuss prospects for the coming season. After I had made my pitch, Calvin began by saying that he offered all the services that I did—"but just did a better job at them." Later in the meeting, when I had joked about bringing some literature with me to support my aging memory, he quipped that "it's not his memory, which suffers from age but rather his eyesight." As a matter of fact, my corrected eyesight is still at least 20/20, even now twenty-two years later. A rumor that my customers were getting by with less insecticide use than his had been circulating for several years. Although I had not mentioned or heard this at the meeting, I believed that his unfriendly statement may have been a reaction to that often-heard rumor.

My strength and interests had always been in collecting data and in crunching numbers, or in reviewing such efforts of others, to learn more of nature's secrets in order to improve crop management practices. I was able to explain the results of such efforts in meaningful and useful ways to potential users of the technology. However, I was never well-equipped to deal successfully with the many kinds of personal interactions required in the wheeling and dealing of business management and promotion. I made needed progress in some of these areas over the years, but much always remained to be desired.

My continual frustration with campus politics and policies at Nicholls, now Nicholls State University (NSU), made, I believed, by politically inspired administrators, often seeking favors rather than excellence, and giving priority to their own welfare above that of the institution they led, sometimes resulted in my overreacting to events. For example, Dr. Bob Falgoust was a young Terrebonne Parish native with a PhD in horticulture who had become head of the agriculture department. He was popular with the students who flocked into his courses. On an occasion in 1981 when he had represented the university at an out-of-town conference dealing with pest control and the future of sugar cane, he was interviewed by the *Daily Comet* upon his return. The newspaper published his statements that included some remarks indicating that he had missed a few points made at the meeting due to a lack of familiarity with an entomological subject. I considered this a poor reflection on the university and on me for the fact that I had not been sent to such a meeting although I was the only entomologist on the NSU campus. Several days later, the *Daily Comet* published my letter to the editor under the heading "Wants correct news from Nicholls." I sent personal copies to the university president, provost, vice president of academic affairs, and dean of life sciences and technology, which only served to allow me to vent my anger. I lambasted Dr. Falgoust's statements with such phrases as "gross ignorance," "flagrantly erroneous," and "those who dabble in areas outside their own specialties." I don't think that Bob ever appreciated my reason for being angry. Of course, in truth, I was also envious of his favored status as a "good old boy" with a glad hand for everyone. He continued to be promoted by the university as its representative to the world in all matters of agriculture and sugar cane, often attending conferences on sugar cane research in Central and South America, sometimes accompanied by other university administrators, for what benefit to sugar cane culture I could never imagine. One objective might have been to attract foreign students to the agriculture program at NSU, which did include some foreign students. However, the total numbers of students in agriculture there declined steadily from the late 1960s through the 1990s.

Bob Williams was training on-the-job for a store management position with K(atz)&B(estoff) Drugs. Jan and Bob were married April 7, 1981, at an LDS chapel in north Louisiana. We attended the wedding. Janice kept Jan's children while they honeymooned for three days in Jackson, Mississippi. We had played major roles in helping to rear our first three grandchildren until Holly was six, Denny five, and Heather going on three years old. Two months after their marriage, Bob moved his wife and family to the small town of Tallulah in north Louisiana, about five hours away. They moved back to south Louisiana for a time, and then to Raleigh, North Carolina, for eight years before moving back to north Louisiana. Their family grew in size with the added births of Heidi Marie, Amanda Louise, and John Henry.

October 6-8, 1981, I attended the Second Inter-American Sugar Cane Seminar on Insect and Rodent Pests at the Florida International University in Miami, Florida, to which I was invited to speak as a panel member on the subject of "Practical Management of the Sugarcane Borer." Other panel members were Dr. Reagan (LSU), Dr. David Hall (Florida), and Dr. Miguel Abarca (Mexico). I showed, with colored maps depicting weekly scouting reports, how a pest management program was actually carried out during a growing season on three different farms. On one farm, insecticide was applied on two different dates to a total of only 23 percent of the crop. On a second farm, treatments were made on nine different dates with all fields being treated at least once while others were sprayed two, three, or four times for an average of almost three overall insecticide applications. On a third farm, weekly SCB infestations never exceeded 5 percent, and the farm was not sprayed at all. In summary, I said that this IPM program integrates natural control factors, cultural factors, varietal resistance, and chemical insecticides into a unified program in which as much benefit as possible is obtained from each control factor and in which insecticide is used only as a last resort. I emphasized that the program provides effective insect control with the advantages of keeping expenses to a minimum, delaying or preventing the development of resistance to insecticides and reducing the magnitude of environmental problems sometimes caused by the adverse effects of pesticides on nontarget organisms. I suggested that this program could serve as a model for SCB control in other countries and areas where stalk borers are a problem. A lengthy question and answer period followed from which I left feeling very good about the program that I had developed in cooperation with Sess Hensley and with considerable inspiration from Dale Newsom and the help of assistants and students.

In 1982, I was appointed by the NSU president to chair a self-study committee of eleven faculty members of the College of Life Sciences and Technology to assess the status of the college and its five departments, their programs, projected plans, needs, conformity with guidelines of the Southern Association of Colleges and Universities, and to present our report duplicated and bound in book form by late 1983. We met weekly for more than a year before presenting our report early in 1984. The report was critical of past performances in some areas, ambitious in others, and specific in its numerous recommendations. Dr. Alice Pecoraro, who served both as editor and committee member, wrote a nice letter of appreciation to me. She was a very positive person by nature who went on later to become a vice president of the university under Dr. Ayo.

Early in 1983, I received a call from the editor of *Sugar Journal*, a technical publication that informs the international sugar industry about cane and beet sugar production, asking permission to spend some time with me in the field. I happily obliged. He spent half a day interviewing and accompanying me as I scouted cane fields. In the March 1983 edition of his journal, there appeared a two-page article entitled "Pest Control for Successful Sugar Cane Production," including a picture of me in sweaty khakis standing beside my pickup truck while taking field maps from a box on the back rack of a three-wheel Honda ATV. His article briefly summarized some events of my career to this time and the development of IPM in sugar cane.

In the matter of promoting my business, most new clients were obtained by word-of-mouth recommendations from satisfied customers. Seeking out and visiting potential customers was sometimes but rarely helpful. Nevertheless, I routinely mailed letters to clients and prospective customers describing my services and prices and informing them of news items from time to time that might be of benefit. By the spring of 1983, my assistant, John Sturgis, had worked with me for five summers. He had graduated from Nicholls State in agriculture, taken my entomology course, and become a dependable and experienced field scout. I felt that by letting him work more independently, we could cover more acres weekly, and I could increase my territory significantly to perhaps thirty thousand acres. Thus my spring letter to growers that year was more lengthy than usual in telling why they needed my service, why I was qualified to do it, and what it would cost. Following are four paragraphs from that letter:

> It is obvious from these facts that I have been positioned for the last twenty-six years on the leading edge of change whenever sugar cane insect pest management methods were involved. This often has not won me friends but generally has benefited the industry. Salesmanship and public relations have never been my strong points, and my business has often suffered because of this. However, I have no regrets for having spent some of my best years in cane country, and I plan to spend the rest of them here.
>
> Regarding my business, there is no better consulting service than mine available anywhere in Louisiana, or in the world for that matter, in the subject area of sugar cane pest management. I have a unique combination of research and practical experience, which eminently qualifies me to help

you make the right decisions about insect control in sugar cane. I am robustly healthy, inherently tough, and thoroughly accustomed to walking in rainy and muddy fields from daylight to dark. I am prepared to do this again for you this year weekly from June till September, or as long as needed, to determine if, when, where, and what you may need to spray for borers.

My fee, which covers the entire season, will remain the same as for the last two years, i.e., $1.75 per acre. This entitles you to a weekly report of infestation conditions on your farm, and to outline maps showing exactly where and when to treat, if and when treatment is needed. I and my assistant will be on Honda 3-wheelers during wet weather so that your field roads and headlands will not be cut up.

If you plan to use my service again or for the first time this year, or if you would like to talk about it before deciding, please complete, sign, and return the enclosed self-addressed postcard.

In 1982, it occurred to me that most of the fire ant studies in sugar cane, which I had initiated earlier at LSU and which had been continued by my successors there, had utilized pit traps exclusively to collect foraging ants for later enumeration and identification. As I spent many hours weekly during the summer months scouting cane fields and often observing foraging ants on the plants, I thought how easy it would be, with just a little extra effort, to collect these ants from the plants on which their SCB prey were present and where the two were more closely associated than on the soil surface where pit traps commonly are set. I recruited three other biology faculty members, two of whom

(Calvin Viator and Paul Templet) scouted cane fields as I did during the summers, to collect foraging ants weekly from cane plants in eighty different fields, forty in 1982 and forty in 1983, scattered over the sugar cane area of the state. These were all fields for which we also had records of recent soil analyses done for our customers. This effort provided a gold mine of new information that I presented in February 1984 at a meeting of the American Society of Sugar Cane Technologists (ASSCT) in Baton Rouge and that we later published (*Jour. Am. Soc. Sugar Cane Technol.* 7:5-14, 1987) after considerable hassling from an LSU entomologist.

In April 1984, Janice flew to Mexico City with Nancy and the E. D. White Catholic High School Band, in which Nancy played the flute and piccolo for five years. She was the prettiest and most talented tryout for drum major her junior year, in the opinion of her parents. However, her father had not been as active as some parents in supporting school activities with his time or money and certainly had not spent much time hobnobbing or politicking with the powers that be. Nancy was elected Miss Teen Thibodaux the following year (1985) and had been a favorite teen model for several years previously in the local stores. She had dated for more than two years a top student and baseball player from Central Lafourche High School who later attended Nicholls State University on an academic scholarship and graduated there with a major in mathematics.

Frank Verdun was a personable young man, and they always appeared happy together. They should have been married before Joshua's birth. However, if they had been, school rules would not have permitted Nancy to graduate with her class that she was determined to do. Following her graduation in May, she and Frank went to look at affordable student housing on campus. Frank wanted to get married then, but Nancy didn't like the housing situation and wanted to attend college. So marriage was postponed; and her first child, Joshua Joseph, was born on Labor Day, September 2, 1985, at 5:00 AM with Frank standing by shouting "Hot damn, it's a boy!" Frank and Nancy had been fishing in the bayou from our backyard the day before when Nancy came in the house and told her mother that she thought it was time to go to the hospital. Since grandmothers and whole families were not allowed in the room at birth as they are today, MeeMaw heard his first cry from just outside the door while standing in the hall.

After graduation, Frank took a job in Texas where Nancy and Josh visited him to look at apartments and child care options. They were still talking marriage. Two years later, Frank came to visit prior to their scheduled wedding. The bridesmaids had bought their dresses, the photographer had been paid, and all was set for a church wedding. Problem was that Frank had changed his mind. I believed that this should end any further contact between us and Frank, and he was no longer welcome to visit. However, Nancy took Josh to see Frank one more time after the wedding was called off, and Josh began to miss his daddy with a passion. Frank wanted to send monthly child support checks in return for visiting rights, but I didn't want to see Frank at all. From the time Josh could walk until he was school age, I often spent time with him in the yard throwing a ball or catching grasshoppers when he would sometimes look up to see an airplane flying high overhead and exclaim while pointing, "My daddy lives in the sky." This scene and others finally convinced me to permit Frank to visit and to accept his monthly child support, which helped pay Josh's tuition at nursery school and later at Saint Joseph Elementary School.

In August 1985, the *Daily Comet* published an article titled "Distinguished Professors Named," picturing Dr. Ayo and Dr. Long standing together between two other NSU professors and stating that Dr. Ayo has honored three members of the faculty as distinguished service professors in recognition of outstanding performance in the classroom, research, and community service. And the following month, Long Pest Management Inc. was listed in *Ag Consultant and Fieldman* magazine as one of the "Top 50 Agricultural Consulting Firms in the United States."

Dr. T. E. Reagan had earned his MS degree under Dr. Hensley at LSU before going on to North Carolina State University for his PhD and then returning to Louisiana as the LSU entomologist in charge of sugar cane insect research. Dr. Reagan had attended my presentation in 1984 and had been particularly impressed by our conclusions that associated greater fire ant numbers with heavier textured soils. He set out immediately to see if he and his co-workers could confirm these results. They did so in a season and hastened to publish in 1985. We submitted our paper, "Abundance of Foraging Ant Predators of the Sugarcane Borer in Relation to Soil and Other Factors," also in 1985 for publication in the *J. Am. Soc. Sug. Cane Technol.* In June 1986, the editor, Dr. Freddie Martin, wrote that it had been accepted for publication. However, five months later, Freddie wrote to say that publication of our paper now was being questioned on grounds of its significance in view of the fact that Dr. Reagan had already published such results. This resulted in an ugly exchange of correspondence between me and Reagan. Ultimately, the priority of our work was established, in spite of such confusion, and our paper was finally published in 1987. Highlights of our work had appeared in local newspapers in 1984 because of its practical interest in answers to such questions as: Why are SCBs more troublesome in the light sandy than in the heavier clay soils? How has the species composition of ant populations in cane fields changed during the last quarter century? How are ant populations affected by currently used insecticides for SCB control?

In 1986, I believed that I was the most competent person in the entire world to scout SCB populations and recommend measures for their control. Thus, it came as a serious and humbling shock, causing me many sleepless nights when I lost a customer (Denis Thibodeaux & Sons Inc.) by making a recommendation that later turned out to be unjustified. In early August of that year, I had recommended a treatment for SCBs on some of their properties. The next week, I and my newly employed scout examined the fields and, finding relatively little infestation, reported to Denis that everything looked good. To my surprise, Denis told me that he had never gotten around to treating as recommended and, of course, was glad now that he had not. So we went back and looked again, combing the fields for most of a day to be as certain as possible about what was going on. We concluded that there was modest but sufficient SCB infestation above our treatment threshold to justify again recommending treatment although there was also evidence of active fire ant predation of SCBs. I went back and took Denis into his fields with me to convince him that he needed to treat to be on the safe side and that this one application should be sufficient for the season. Denis saw the little SCBs but did not agree that so small and so few borers could justify the cost of treating. I invited him to call in another consultant for a possible alternative recommendation, which he did and which confirmed for him his belief that treatment was not justified. I was so upset by the situation that I continued to scout his property weekly through the rest of the critical period and into the harvest season. When I finally gave it up, I had to conclude that I looked like a fool and that Denis indeed had saved money by not treating.

The fact that Denis lost big money the next year on another plantation for following his new consultant's advice was not much comfort to me. I had lost the trust of a good customer and friend. However, these experiences emphasized, in the most forceful possible manner, the importance of good judgment in making insect control decisions in field crops, the complexity of factors that may impact such decisions, and the fact that no one (me included) can forever carry a batting average of one thousand. Some factors that may influence these decisions in Louisiana sugar cane include: numbers and developmental stages of pests present, age and condition of the crop, developmental stage of the crop, the crop variety being grown, abundance and activity of beneficial predators, soil type, weed conditions, present and future weather conditions, costs of treatment, and our inability to predict with complete accuracy the effects of any one or all of these interacting factors upon crop yield. Even knowledge of the past history of pest problems in the particular field(s) may sometimes help tip the balance of judgment in one direction or another.

Dr. Curt Rose and Dr. Alva Harris were marine biologists at NSU with considerable expertise in their fields. They were first to discover a large "dead zone" in Gulf waters surrounding the mouth of the Mississippi River. On this basis, they obtained significant research funding from federal government sources to enlarge and pursue these studies. Their work received a considerable amount of deserved publicity and was in an area, namely scientific research, with which the NSU administration had little experience. Some believed that the administrators were better at politics than administration and sometimes more interested in access to personal benefits than in the best interests of faculty or university. It was said that on one occasion, various members of the administration—including the president, his vice presidents, and the college farm manager—all had attended research planning meetings and scientific conferences while the principal research scientists were unable to attend and the university brass were entertained at the expense of the marine research grant funds. Milk, meat, and produce from the college farm were "perks" said to be common in the president's dining room and in the homes of friends. The marine research laboratory, located an hour away on the Gulf, was said to be a favorite place for administrators and their friends to fish and party. These stories were common subjects of conversation among the faculty.

This atmosphere, combined with campus politics under an administration that many believed to favor promoting loyalty above qualification, resulted in polarization of the faculty into two groups—one in support of and one in opposition to the administration. Many faculty members, including me, who had never belonged to a union in their lives, joined the AFL-CIO Teachers Union. Our objectives were either to influence for good or to oust, if possible, the university president, Dr. Ayo, who had succeeded Dr. Galliano in 1983. These efforts were abandoned, as was much faculty membership in the union, upon discovery that our union contact was actually keeping the university administration informed of our activities.

Governor Roemer received a letter from me, dated January 12, 1989, gently scolding him for prematurely endorsing the appointment of Dr. Ayo as president of NSU. I wrote the governor again in May to encourage his support of the investigator general in the investigation of practices at NSU. At a faculty senate meeting in September 1989, psychology professor Gary Ross-Reynolds said, "We owe it to the president to get a feel on how the faculty feels." Ayo had now been president for six years. However, the senate agreed to wait until after the current semester to vote on the referendum of confidence in Ayo as president. I was then a member of the senate which voted 10-4 that day to give the faculty a chance to vote next January on confidence or lack thereof in the president. At the end of the meeting, senate president Tom Butler opened the floor for members of the audience to speak, at which student government president Rocky Capello and assistant professor Laura Badeaux expressed their outrage for the "lack of objectivity" and the "abuse of parliamentary law" shown by senators. The following January, a majority of 66 percent of the faculty voted no confidence in Dr. Ayo's administration, not surprising in view of the list of rumored abuses of authority or lack of supervision associated with his tenures as vice president, provost, and president.

Competition between agricultural consultants and LSU extension and research personnel for respect and for opportunities to serve farmers at the grassroots level was increasing and sometimes caused concern and confusion in the state's sugar industry. Attempting to shed light on the issue and to promote better communication between consultants and research/extension personnel, the American Sugar Cane League organized a meeting of grower support groups for March 3, 1989, at the Sheraton Motel in Thibodaux. It was well-attended by Extension Service personnel, including county agents, LSU and USDA researchers, and agricultural consultants working with sugar cane. A number of people were asked to speak during the meeting that lasted most of the day. I used about fifteen minutes to say that the relationship between farmers and consultants was like that between patients and their doctor, and the idea that extension and research personnel could service all these individual needs was no more reasonable than to believe that the faculties of the medical schools could take care of all the patients in the state. Six days later, I received a phone call from Dr. Rouse Caffey, chancellor of LSU, expressing his unhappiness with my remarks and stating that we needed to talk. I responded that I would be happy to do so at our mutual convenience. He never followed up on the matter.

John Sturgis had departed from Long Pest Management Inc. in 1985 to accept employment as a game warden with the state of Louisiana, which would give him better medical coverage and retirement benefits and would not require him to move back and forth annually between Florida and Louisiana. I then hired Steve Hoak, another Nicholls State student in agriculture who earned his BS degree the next year (1986) and began to work with me during the field-scouting season from May through September. I helped Steve to get employment during the rest of the year with Glades Crop Care Inc. in Florida as I had done for John. Glades had some seasonal needs for additional field scouts, and my needs for additional field help were mostly seasonal. Steve was a hard worker and very interested in the work, about which he pursued learning both by reading and attending meetings at every opportunity. He was a practical, down-to-earth big strong fellow who mixed and got along well with customers. He steadily gained their respect as they saw his dedication to the work and his willingness to endure all kinds of mud, rain, and heat to get the job done. Daniel was in high school then and helped out with winter soil sampling and other chores on a part-time basis by which he earned some spending money.

Dr. Wray Birchfield had studied and published extensively on the subject of nematodes as pests of sugar cane during nearly four decades but was never able to develop widely acceptable and practicable control recommendations. Nevertheless, when I observed sugar cane heavily stressed in a manner possibly suggestive of nematodes, I did some assays of soil samples in 1986 which confirmed a great abundance of plant parasitic nematodes in the soil around these sugar cane roots. With the cooperation of Herman Waguespack, an agronomist with the American Sugar Cane League (ASCL), we conducted field tests for three years (1987-89) to determine possible benefits from nematode control by soil treatment with a leading nematicide chemical. Initial results (*J. Am. Soc. Sug. Cane Technol.* 10:79-84) were sufficiently encouraging for the ASCL to fund further research under my direction. I attended a nematode identification short course at Clemson University during December 1989 and January 1990. Subsequent studies continued to show promise, but not enough in terms of crop yield benefits to justify the costs of treatments. And so after two more years, funding of the project was discontinued. Meanwhile, the fact that an entomologist at

NSU was dabbling in a field of research more commonly considered the domain of pathologists it seemed stirred some interest at LSU in further research on the subject there. After a few more years, that interest also waned, and no practical chemical control recommendations were ever developed for nematode control on sugar cane.

Daniel was always a bright, witty, highly social, aggressive, strong-willed, athletic, handsome child with an unusually high energy level. As coach of his Biddy Basketball team, I had benched him and a much larger teammate during the entire first half of a game for fighting during the pregame warm-up. During high school days, he often got home from football practice to run full speed around the house for half an hour before supper while his teammates in the neighborhood preferred to rest rather than swim with him in the pool. One afternoon, I came home to find him rushing down the bank to dive into an alligator-infested bayou instead of swimming in his own twenty-by-forty-foot pool. In high school, at a weight of about one hundred sixty-five pounds, he whipped a two-hundred-pound tackle on the football team to a state of helplessness for intentionally throwing a bottle that cracked the windshield of our car. As a child, he played on city all-star basketball teams almost every year. As an eighteen-year-old, he played for the state championship on an American Legion baseball team. Through elementary and high school, he was always a standout athlete, but was never a favorite with the school coaches, and ran with a wild group of friends whose top priority was partying. In some ways, he was living a life that I could only have dreamed of at his age, which may have made it difficult for me to be as strict as I should have been. In spite of his abilities and our attempts to encourage his interest in books and homework, even to the extent of hiring tutors for some of his courses, he was never academically inclined and often turned the tutoring sessions into a comedian's performance.

During the late seventies and eighties, many kids in both the private and public schools drank and smoked pot, and for some, this led to addictions with serious consequences. Shielding them altogether from these influences would have required almost complete isolation from their peers. Nancy and Dan were strong-willed, outgoing, and very socially inclined. Janice attended LDS church services regularly and usually was accompanied by Jan, Jan's children, and Nancy. Because Sunday services at the Presbyterian Church were only two hours instead of three, Dan commonly preferred attending church with his daddy at the Presbyterian Church. There were not many young people in either congregation. We had scripture reading with family prayers on a hit-or-miss basis, which became increasingly difficult as I became more and more involved with school and business activities and as the children became teenagers. Janice worked hard at keeping house and feeding the family in our large two-story home where teenage rebellion emerged and seemed to grow. Enforcing midnight curfews became a losing battle, and few, if any of their acquaintances, lived by such rules.

Carrie Joseph was a dependable and helpful black lady, who began working for us in 1970 when she was in her early twenties and Dan was a baby. She had two girls of her own and often confided to Janice about how much they "vexed her nerves." She continued with us for more than thirty years working one day per week most of the time, usually on Fridays, tidying up, cleaning, washing, and sometimes ironing. She occasionally let it be known that her children weren't the only ones who "vexed her nerves" and she couldn't understand why ours weren't more helpful at home, why they didn't appreciate more what they had, and why they didn't show more respect for their parents.

We had just celebrated Nancy's twentieth birthday. Dan was less than a month away from his eighteenth, and Josh was twenty-seven months old. Janice's parents and Aunt Chloe were looking forward to seeing baby Josh for the first time and having us all at their house for Christmas in Raleigh, North Carolina. We had planned to be there all together for a family Christmas. For this reason, we had not put up a tree or decorated the house although the date was December 22, 1987. For Nancy and Dan, the prospect of a seventeen-hour drive going and coming and sleeping on a couch or floor mat for several days in a town far from their many friends, who would be partying in Thibodaux, was not appealing. Therefore, on the night before our departure, our two youngest children announced, when asked why they were not packing, that they were not going. Janice and I both were surprised and unhappy with such a turn of events. We didn't like the idea of Nancy and Dan having the unchaperoned run of the house for several days with their friends, but we were not going to disappoint the family in Raleigh to satisfy the whims of our rebellious children. And so we left the next morning without them. Nancy says that when they got up the next day, we were gone and that we left them without groceries or money, which is certainly true. Janice cried about it all the way to Raleigh. Upon our return home, a Christmas tree had been put up and decorated as well as the house. Their friends had brought food, and a Christmas Eve dinner had been held. Frank had come and played Santa Clause for Josh.

November 29, 1988, six months after his high school graduation, I took Dan to the bus station to catch a bus to the army barracks in New Orleans from which he would fly to Ft. Sill, Oklahoma, for basic training. And so two months before his nineteenth birthday, he would miss Christmas at home. He had already missed two other Christmases at home to attend baseball camps. During his childhood, I had spent many hours with him throwing footballs and baseballs and swinging a ball on the end of a rope for batting practice. But I always had felt that somehow he wasn't getting from me all that he needed, and I was no expert in athletic matters. It sometimes seemed that his intense interest in sports was based as much on his particular social needs as on his desire to master the art of the game for its own sake. I concluded that in either event, mastering the art of the sport should help him to achieve his goals. And so during high school for two other Christmases, he had spent two weeks each at the Show-Me baseball camp in Missouri and the Dodgers winter camp in Florida learning from the pros. Janice tells me now that it was all a great mistake and that, if he would admit it, he would say that I had made him feel unwanted at home. May he and God both forgive me if that is true.

Dr. Dale Newsom suffered a heart attack and suddenly passed away in 1987. He was an inspiring scientist and leader for whom I had the greatest admiration and respect and whose influence had been as helpful to me as that of anyone I had ever known. The following year, the Louisiana Agriculture Consultants Association voted to honor him by election to the Louisiana Agriculture Hall of Fame. I was appointed to present the eulogy and to make the posthumous presentation to his widow at our winter meeting in 1989. It was an assignment that I approached with reverence and that I almost failed to complete by choking with emotion at the podium.

In early March 1989, President Ayo appointed me to serve on a search committee for a new vice president of Academic Affairs. Our committee worked hard reviewing applications, calling references, and interviewing several top candidates. We finally selected Dr. Al Etheridge from the University of Arkansas at Monticello, and he was confirmed by the board in July of that year. Many faculty members considered him a much-needed breath of fresh air in an administration with many needs. However, after four or five years, he left to assume the presidency of a university in Pennsylvania.

Nancy graduated in May 1989 from Nicholls State University with a BS in Business Management, after which Janice drove with her and Josh to Raleigh, North Carolina, to visit family in early

June before returning two weeks later with two more grandchildren, Heather and Heidi. After entertaining Heather and Heidi for a month, she took them back to Raleigh in late July and returned home in mid-August.

The summer of 1989 was typically hot and wet with frequent thunderstorms. Field scouting continued in all types of weather unless lightning forced us to take temporary cover in our trucks. I took care of most of the office work that was done, and Steve and I shared about equally in the fieldwork. He arrived from Florida in May to help finish preparing field maps and, after a hard summer's work, returned to Florida in September.

Janice's interests were confined mostly to the children, housekeeping, gardening, and shopping. My workaholic habits continued to deny her the time and attention from me that she deserved and needed. Consequently, she spent more than a normal amount of time and money shopping and traveling to and from North Carolina to visit family or to take children or grandchildren there. I usually accompanied her and the children on a beach trip annually but often left them to get back to work before the week was over. This behavior of mine was frustrating to her, and my inability to get ahead financially, in spite of working two jobs as hard as I could, was frustrating to me. Getting ahead to me meant setting aside and investing money for the future. For Janice, it meant having more and living better now since we might not be here tomorrow. She often accompanied me to several professional meetings and seminars annually at which there were special activities for the ladies while the men attended meetings. And so the seventies and eighties, which seemed to drag then, passed more rapidly than now seems possible. "For what is your life? It is even a vapour, that appeareth for a little time, and then vanisheth away" (James 4:14).

15

IPM AND THE REAGAN PROBLEM

Early in 1993, I received a phone call from Dr. Charlie Richard, senior agronomist with the ASCL. Charlie said, "Henry, are you aware that Dr. Gene Reagan has been selected as the nominee of the Southeastern Branch of the Entomological Society of America (ESA) for the J. E. Bussart Memorial Award in recognition of his important contributions to economic entomology? I had not heard the news but found in my mail a notice of the approaching (March 7-10) Sixty-seventh Annual Meeting of the ESA that contained the announcement about Gene Reagan. I notified Sess Hensley, who was both surprised and angered, and then began to compose a letter to the chairman of the awards committee. I believed that Gene had tried to block publication of my ant research almost a decade earlier. He had been a pain in the ass in his efforts to prevent the use of Azodrin insecticide by the industry in the early eighties. And now he seemed to be seeking to take credit for the accomplishments of two respected entomologists who, for twenty years before his arrival, had done the work of developing IPM in sugar cane and greatly improving related agricultural practices.

I wrote the following letter, which Sess and I collaborated on and both signed before mailing February 15, 1993, to Dr. John R.

McVay, chair, Member Awards Committee, Southeastern Branch, ESA, 207 Extension Hall, Auburn University, Alabama 36849:

> Dear Dr. McVay:
>
> We, the signatories and longtime members of the Entomological Society of America, Southeastern Branch, write this letter in formal protest of the recent nomination of Dr. T. E. "Gene" Reagan as our branch's nominee for the J. E. Bussart Memorial Award. We regret having to ask for time in your busy schedule and that of other members who may become involved in deliberating this matter. However, when the facts are known, we believe that some corrective action will be deemed appropriate in the interests of ethics and justice.
>
> Page 8 of the program for the Sixty-seventh Annual Meeting of the Branch summarizes Dr. Reagan's qualifications for this nomination. Following are excerpts from this summary:
>
> 1. "One can perhaps best summarize these contributions by pointing out that, since he came to LSU in 1977, the number of applications of insecticide per season for control of sugarcane borer in Louisiana has been reduced from 3.5 to less than one per year."
> 2. "His IPM program in sugar cane emphasizes a balance of cultural, biological, and chemical control tactics . . ."
> 3. Dr. Reagan developed a new biological assay for evaluating cultivar resistance to the sugarcane borer, thus providing "entomological input into the sugar cane breeding program, resulting in the

release of a new variety with more borer resistance than any other commercially grown variety."
4. Dr. Reagan has developed a new IPM model, which enables him to forecast the impact of variety releases, weather, soil type, weed management, and predator ecology on sugarcane borer populations.
5. "Gene's . . . contributions are outstanding, and he personifies the qualities the J. E. Bussart Award is designed to recognize."

Dr. Reagan joined the LSU faculty in 1977, approximately sixteen years ago. Some of us, who have been active in sugarcane borer (SCB) research and/or consulting for more than thirty years, believe that a number of facts demand response to the above-quoted excerpts. The facts which follow should be considered in relation to the numbered excerpts above.

Long and Hensley, writing in volume 17, page 168 of the *Annual Review of Entomology* (1972) stated "During the past decade, several changes in control procedures recommended to Louisiana cane growers have resulted in reducing the maximum number of insecticide applications required annually for control of *D. saccharalis* from twelve to three." The first truly effective method for controlling the SCB had been made available to Louisiana cane farmers in 1959 when the use of four biweekly applications of endrin granules was recommended by both LSU and USDA entomologists (*Sug. Bul.* 37(14):167-8). The discontinuance of first generation SCB control at that time eliminated the possible need of four early season insecticide applications (Hensley, Long et al., *J. Econ. Entomol.* 56(3):407-9). The introduction of the highly effective insecticide endrin, which did not require weekly applications (Long, Hensley et al., *J. Econ. Entomol.* 52(5):821-24), further reduced the maximum number of treatments required from eight to three and ended the ineffective practices of weekly dusting with ryania in combination with annual releases of laboratory-reared egg parasites. The significant facts here are confirmed by comparing the published recommendations for SCB control in 1956 (*Sug. Bul.* 34(13):191-2) with those of 1962 (*Sug. Bul.* 40(19):210-11).

The year when insecticides were first recommended in Louisiana sugar cane as part of an IPM system was 1964 (Long and Concienne, *Sug. Bul.* 42(15):183-4). These recommendations were based upon research conducted cooperatively by Long and Hensley (and their coworkers) during the previous seven years. Emphasis was on control by cultural practices, varietal resistance, conservation of predators, and the judicious use of insecticides only when and where necessary as determined by weekly field scouting. These practices (particularly the latter) were quickly adopted and resulted in further reducing the required number of insecticide applications from three to two or less (Hensley, *Sug. Jour.* 44(2):18, 1981). The average number of applications required annually after 1964 was no more than two and often less.

The impact of these first highly effective SCB control practices on the Louisiana sugar industry is reflected by data on per acre sugar yields from the office of the ASCL, Thibodaux, Louisiana. From these data, it can be shown that the average per acre yield of sugar increased from 3,195 lbs, during nine years before 1959 when the use of synthetic organic insecticides was first recommended in Louisiana, to 4,280 lbs in the decade immediately following that year. This 34 percent increase in sugar production between the fifties and sixties resulted largely from better insect control and, to a lesser extent, from improved crop varieties.

The critical period for controlling the SCB was further defined by Long and co-workers who demonstrated decreasing crop responses to SCB control with increasing size and maturity of sugar cane plants (*J. Econ. Entomol.* 57(3):350-53, 1964; *Sug. Bul.* 60(21):9-11, 1982). This led to a reduced frequency of late-season insecticide treatments. The occurrence of a weak larval diapause, described by Katiyar and Long (*J. Econ. Entomol.* 54(2):285-87, 1961), contributes to declining SCB infestation pressure in late summer and fall and also suggests less need for concern about late-season population growth and crop losses to this insect.

Long and Hensley and coworkers were first to recognize in the late 1950s the beneficial role of the red imported fire ant in sugar cane fields (*Sug. Bul.* 37(5):62-63, 1958; *J. Econ. Entomol.* 54(1):146-149, 1961). Their studies were cited by Rachel Carson in her book, Silent Spring. Since that time, all recommended chemical control practices have been designed to have minimum adverse effect on ant populations. Further reductions in insecticide use have accompanied increasing entrenchment of the red imported fire ant as the dominant insect predator in our cane fields during the years from 1960 to 1984 (Long et al., *J. Am. Soc. Sug. Cane Technol.* 7:5-14, 1987).

The increasing effectiveness of ants as predators was facilitated in part by discontinuing in 1965 the general use of chlordane in the planting furrow to control soil micro arthropods. This change in recommendations (compare *Sug. Bul.* 42(15):183-4, 1964 with *Sug. Bul.* 43(17):204-5, 1965) was based on studies by Long and coworkers (*J. Econ. Entomol.* 60(3):623-9, 1967), which indicated that early stimulation of plant growth by chlordane was not accompanied by yield responses at harvest and that the springtails and other microarthropods did not significantly affect sugar cane yields.

Increased planting of varieties with some resistance to the SCB has also contributed significantly to reduced insecticide use. In 1957, about 55 percent of the Louisiana cane acreage was occupied by the variety CP44-101 (*Sug. Bul.* 45(3):389, 1966), which was highly susceptible to the SCB. Hensley and Long demonstrated a wide range in yield loss to the SCB among commercial sugar cane varieties in Louisiana in 1969 (*J. Econ. Entomol.* 62(3):620-22). Their data indicated that as much as $50 per acre could be saved annually by utilizing SCB resistant varieties. Hensley et al. stated in 1977 that about 50 percent of the Louisiana sugar cane crop was devoted at that time to production of NCo310, CP52-68 and L62-96, which are varieties with useful levels of resistance to the SCB (*Proc. Int. Soc. Sug. Cane Technol.* 16(1):517-22). In 1991, varieties CP65-357, CP70-321, and CP72-370, rated as moderately to highly resistant, made up 86 percent of the state's cane acreage (*Sug. Bul.* 70(11):22, Table 2; *Insect Control Guide*, LSU Coop. Ext. Ser., p. 128, 1992). These facts show that increasing advantage has been taken of varietal resistance to the SCB during the last thirty-five years. Resistance of these varieties to the SCB was determined by a conventional method of comparing differences in percentages of joints bored among candidate varieties and did not involve the use of Dr. Reagan's "new biological assay." We know of no commercial variety for which resistance was determined by this assay.

In comparing a relatively resistant (NCo310) and susceptible (CP44-101) variety, Kyle and Hensley concluded that ovipositional preference was not important and that varietal resistance was due mainly to mortality of young larvae before they penetrated stalks (*Proc. La. Acad. Sci.* 33:55-67). Coburn and Hensley found that a combination of leaf sheath appression to stalks and rind hardness of immature internodes were most responsible for SCB resistance in sugar cane variety NCo310 (*Proc. Int. Soc. Sug. Cane Technol.* 14:440-44, 1972). Martin, Richard, and Hensley subsequently showed that penetrometer measurements of rind hardness could be used to determine partial resistance to the SCB (*Environ. Entomol.* 4(5):687-8, 1975). Pan and Hensley showed that it is impractical to screen for SCB resistance in seedling sugar cane plants by infesting them with newly hatched larvae (*Environ. Entomol.* 2(1):149-54). White and Hensley developed new techniques to quantify the effects of SCB on sugar cane quality (*Field Crops Res.* 15:341-48, 1987). White et al. (in press) have documented and registered five sugar cane clones with SCB-resistant germ plasm. Bessin,

Reagan, et al. described a moth production index for evaluating sugar cane cultivars for SCB resistance (*J. Econ. Entomol.* 83(1):221-25, 1990) and suggested improved statistical methods for using percent joints bored to evaluate the resistance of varieties (*J. Am. Soc. Sug. Cane Technol.* 10:8-22, 1990). These studies summarize the significant information on mechanisms of host plant resistance to the SCB and on the evaluation of varietal resistance levels. Obviously, many workers have contributed significantly to this body of knowledge.

Komblas and Long (*J. Econ. Entmol.* 65(2):439-45, 1972) first accurately determined that the seasonal transmission of sugar cane mosaic virus occurred in late winter and spring rather than in summer and fall as previously believed. They demonstrated the impossibility of effectively using a chemical pesticide to achieve practical control of numerous migrating and non-colonizing aphids to significantly suppress mosaic spread.

As a result of the work of Hensley and Long and the widely acclaimed benefits from it, they were invited individually and cooperatively to describe their newly developed IPM system in several prestigious publications. These include:

(1) Long, W. H. 1969. Insecticidal control of moth borers of sugar cane. In *Pests of Sugar Cane*, Elsevier Publishing Co., Amsterdam, pp. 149-161.
(2) Hensley, S. D. 1971. Management of Sugarcane Borer Populations in Louisiana, a Decade of Change. *Entomophaga* 16(1):133-146.
(3) Long, W. H. and S. D. Hensley, 1972. Insect Pests of Sugar Cane. *Ann. Rev. Entomol.* 17:149-176.

(4) Hensley, S. D., H. P. Fanguy and M. J. Giamalva. 1977. The Role of Varietal Resistance in Control of the Sugarcane Borer, *Diatraea saccharalis* (F.), in Louisiana. *Proc. Int. Soc. Sug. Cane Technol.* 16(1):517-522.

(5) Hensley, S. D. 1978. Management of Sugar Cane Insect Pests. *Proc. Int. Symposium on Sugar Cane* in Nigeria, National Cereals Res. Inst., Ibadan.

The rapid adoption of IPM practices in Louisiana sugar cane was realized mainly by the efforts of highly educated, knowledgeable, and experienced agricultural consultants, whose recommendations are and have been followed closely by at least 80 percent of cane growers in this state for the past twenty-five years. None of Dr. Reagan's research has had any discernable impact by changing practices related to sugar production in Louisiana.

It is obvious that the facts stated in the last eleven paragraphs and the statement in excerpt # 1 above cannot both be correct. Use of the words "His IPM program" in Excerpt # 2 is also inaccurate, inappropriate, and dishonest. It is not his IPM program. Furthermore, the program originated by Hensley and Long not only included chemical, cultural, and biological control tactics but also included host plant resistance. This balanced program was in place and recognized internationally (Solomon, 1973) long before Dr. Reagan's arrival at LSU in 1977.

The development of new sugar cane varieties in Louisiana is a team effort involving a large number of players employed by three different agencies, LSU, USDA, and the ASCL. Few people would have the audacity to claim major credit for the release of any new variety. Furthermore, the claim of significant insect resistance in a new variety should be tempered with caution until sizable acreages of that variety have been grown for several years. In view of these considerations, the claim made in excerpt # 3 above seems somewhat presumptuous.

Serious efforts to mathematically model insect phenomena are useful by increasing our understanding of the complex interactions among ecological factors. However, extreme caution must be exercised against overreliance upon the predictive value of such models. If Dr. Reagan's "new IPM model" (*J. Environ. Entomol.* 20(1):252-7) enables him to accurately forecast anything, it has not been obvious to many. In fact, entomologists are still unable to predict future SCB populations with much accuracy. In 1991, a deadheart survey, conducted annually by LSU and USDA entomologists, predicted only average infestations but was followed by one of the most severe seasons of SCB infestation in years. Many other such examples emphasize our present inability to predict future SCB populations with much accuracy.

Dr. Reagan's "new IPM model" actually makes no mention of soil type. The impact of soil type on populations of the SCB was first investigated and understood in 1982-83 by Long et al. (*J. Am. Soc. Sug. Cane Technol.* 7:5-14, 1987). After hearing results of these studies presented in a formal presentation by Long at a LAPPAN meeting in Baton Rouge 3/10/83, at meetings of sugar cane technologists in Baton Rouge 2/2/84 (program and speaker's notes enclosed) and at Fort Walton Beach 6/13/85 (*J. Am. Soc. Sug. Cane Technol.* 6:135), he (Dr. Reagan) co-authored with his students a paper on their 1984-85 studies of the effects of soil types on the fire ant, which was published in 1986. Long et al. first became aware of this through a 11/10/86 letter (enclosed) from the editor of the *J. Am. Soc. Sug. Cane Technol.* informing them that the content of their manuscript (submitted 10/85 and accepted for publication 6/25/86) was very similar to that of one already published in another journal by Reagan and his students (*Agriculture, Ecosystems and Environment* 18:63-71, 1986) and that the significance of the contribution of Long et al. was therefore questioned. Nevertheless, the paper of Long et al. was finally published in 1987.

Facts detailed in the last two paragraphs raise serious doubts of the validity and propriety of the claims made in excerpts # 4 and # 5 above.

We are not opposed in principle to the nomination of Dr. Reagan for an award. However, we are opposed to special recognition of anyone for work which they did not do or for work which actually was done by others. Dr. Reagan has done some good work in determining the relative potentials of insecticides to induce SCB resistance ... and on the destruction of arthropods by mirex ... While these and other contributions of his have provided important details and insights, which enhance understanding of the sugar cane ecosystem, again we reiterate that his research has had no discernable impact on the practices related to sugar production in Louisiana.

Although several television and newspaper articles featuring the research exploits of Dr. Reagan indicate otherwise, the entomological literature clearly and historically establishes the fact that Henry Long and Sess Hensley with coworkers and students, via thorough scientific effort and hard work, are the primary originators of effective IPM in Louisiana sugar cane. Solomon (1973), in addressing the International Congress of Entomology on the subject of ecology in the management of insects, elaborated in detail on this IPM system as an example of practical pest management (*Proc. 14th Int. Cong. Entomol.*, Canberra, Australia, pp. 153-167). Any attempt to assign this honor to another person or persons must either be misguided or judged as unethical.

We respectfully suggest that, in view of these facts, Dr. Gene Reagan should not have been nominated for the Bussart Memorial Award, and we urge the Entomological Society of America to withdraw his nomination from consideration for this prestigious award.

Available reprints of pertinent articles cited above and copies of other documentary material are being sent under separate cover.

Respectfully,

W. Henry Long, PhD, Independent Agricultural Consultant (REAP), Board Certified Entomologist (ESA), Distinguished Service Professor of Biological Sciences, Nicholls State University, P. O. Box 1193, Thibodaux, LA 70302

Sess D. Hensley, PhD, Independent Agricultural Consultant, Professor of Entomology, LSU (retired), Research Leader for Crop Protection, USDA Sugar Cane Field Laboratory (retired), 204 Pendleton Drive, Houma, LA 70360

Copies of this letter were also sent to Dr. Gary A. Herzog, president of the southeastern branch, ESA and to Dr. H. Rouse Caffey, chancellor, LSU Agricultural Center. Sess and I were members of the ESA and had presented papers at ESA meetings, but we did not attend meetings regularly or serve on committees and were not active in ESA politics. Relatively few ESA members worked with sugar cane. Gene Reagan, by comparison, was active in the society, serving on committees and often appearing on meeting programs.

Dr. McVay answered our letter eight days later. He described how he had followed guidelines in forwarding nomination packages to the committee members and said that if there was a single nomination in any category, that individual became the branch nominee unless a consensus of the committee failed to recommend. He said that in reviewing the nomination package submitted on behalf of Dr. Reagan, he had found that it adhered to the guidelines which required that the nomination and subsequent awarding of the "award be made on the basis . . . 'of research work . . . published during the immediately preceding three- or four-year period' . . . and that Dr. Reagan's nomination package adhered to these guidelines and was forwarded to the parent society to be judged on those merits." He went on to say that Dr. Reagan's nomination was made by administrators and his peers at LSU, was well-documented by letters of support from entomologists throughout the United States, and that he could find no provision in the awards guidelines for the protest of a branch nomination although he presumed that the executive committee of the society might overrule committee decisions. In the last portion of his letter, he stated:

> "Your concerns and position are well-presented, and it is quite apparent that the two of you and others built a solid foundation for Sugar Cane IPM in Louisiana . . . it appears that the information shows

how he" (Dr. Reagan) "has worked to compliment that program. It further appears that the wording of the biographical sketch . . . may have been the focal point of your objections. If so, please accept my apologies for any real or perceived injustices . . . I relied on the nominators of the successful awardees to prepare their biographicals. In retrospect, that may have been a mistake, especially in this case . . . I can only hope that you understand my position . . . Unless the president of the Southeastern Branch in conjunction with the Executive Committee finds otherwise, I can see no true justification for stripping Dr. Reagan of his nomination."

We could have pursued the matter further with the ESA but decided this might not be helpful and could be detrimental in the long run. However, Dr. Rouse Caffey, chancellor of LSU, apparently felt that some consultants, I in particular, should be enlightened. He sent a memorandum March 18, 1993, to me, Sess Hensley, Gene Reagan, and five LSU administrators for the purpose of discussing "problems and opportunities in sugar cane entomological research programs," requesting that we meet in his conference room at 1:00 PM Thursday, March 25. I called that I had a teaching conflict at that time. Later another memorandum arrived rescheduling the meeting for Friday, April 2 at 10:00 AM and with four additional names on the committee that now numbered thirteen. From my perspective, there were five committee members in my camp and eight in that of LSU's.

The meeting convened in a conference room of the LSU agricultural center. Dr. Caffey sat at one end of a conference table, flanked by Dr. W. B. Richardson, dean of the college of agriculture, and Dr. Ken Tipton, director of the experiment station. I sat at the opposite end of the table, and arraigned between us on either side were Dr. Denver Loupe, extension service sugar cane specialist; Dr. Bruce Flint, director of the LSU cooperative extension service; Dr. Frank Guillot, head of the department of entomology; Dr. Gene Reagan; Dr. Dale Pollet, extension entomologist; Dr. Sess Hensley; Mr. Charles Melancon, general manager of the ASCL (American Sugar Cane League); Dr. Charles Richard, senior agronomist of the ASCL; and Dr. Ben Legendre, head of the USDA sugar cane field laboratory at Houma, Louisiana. It quickly became clear that Dr. Caffey's objectives were to pour oil on troubled waters and simultaneously maintain the unblemished reputation and leadership of LSU and its faculty in all matters possible. After introductory remarks, Dr. Guillot was asked to talk about Gene Reagan's accomplishments and how he had earned the nomination for the Bussart Award. Dr. Guillot had been at LSU for only a couple of years at that time and had no firsthand knowledge of the history of sugar cane insect research. As he described Gene's achievements, he seemed to be presenting me and Sess in a bad light for having ever questioned the nomination. After listening a few moments, I said, "Frank, you don't know what the hell you're talking about. Just shut up and be quiet!" To this, Dr. Richardson intervened with "We won't tolerate this kind of behavior or disrespect of persons here." To which I replied, "And I won't listen to this kind of crap if it means walking out of here." Dr. Caffey again took charge to calm things down, but the spirit of the meeting remained one of subdued antagonism and ended shortly thereafter.

The personnel of the American Sugar Cane League and USDA at the field laboratory in Houma, Louisiana, apparently recognized our contributions and felt that we had been wronged by Reagan's nomination. Consequently, a few weeks later, Sess and I were invited to be special guests of honor at the 1993 Inter-American Sugar Cane Seminar in Miami, Florida, September 17, at which we were entertained, honored, and presented plaques. My plaque reads as follows:

> The 1993 Inter-American Sugar Cane Seminar on "Sugar Cane and Our Environment" honors and presents this
>
> *Plaque of Recognition*
>
> to
>
> ## *Dr. Henry Long*
>
> In appreciation of his outstanding work in pioneering, developing, and practicing the environmentally sound practice of Integrated Pest Management.
>
> Miami, Florida, September 17th, 1993

The November 1993 issue of the *Sugar Bulletin* carried an article entitled "Louisiana Scientists Honored at Inter-American Sugar Cane Seminar" briefly summarizing each of our careers and with photographs of the honorees.

16

THE NINETIES

There was a swing set in our yard on which Josh often played and on which he was swinging as high as he could after a rain one fall afternoon in October 1990. Janice was in the yard when she heard Josh's cry from the other side of the house and hastened to find him sitting on the grass in front of his swing set and unable to get up. With her hands under each of his armpits, she helped him to his feet and led him to the carport entrance to the house. I opened the door to see Josh standing with arms at his side and sobbing as big tears ran down his face. His forearm on each side was strangely bent outward away from his body. It was obvious at a glance that both arms were broken, like green twigs on a shrub, but without broken skin or protruding bone. When asked what had happened, he sobbed, "I was just swinging and then I was sitting on the ground and it hurts bad." He had been swinging high when he jumped out of the seat and tried to brace his fall with his hands as his feet slipped out from under him on the wet grass. We quickly put him in the car and drove to the hospital after calling Dr. Wes McGee's office to have him meet us there. Three hours later, Wes had set the bones in each arm and with plaster casts on each. Josh had been a brave five-year-old, but the experience of having four bones set without any pain relief had caused him to cry out several times, which caused his MeeMaw to cry most of that time and all the way home. For the next six weeks, she dressed and undressed him, took him to kindergarten, brought him home, and went back to feed him lunch and take him to the bathroom. It was a long six weeks after which Josh had no further interest in swing sets.

Aunt Chloe called in the evening of December 20, 1990, to say that Oscar, Janice's father, was failing fast and in the hospital. Janice flew to Raleigh and called back heartbroken to say that Oscar didn't recognize her. But the next day, he did, and they spent the afternoon visiting and listening to music. After that, he didn't recognize her much anymore but seemed to converse a lot at night, as she sat by his bedside, with persons she could neither see nor hear and who had been dead for a while. He died ten days later at 4:00 AM January 6, 1991, on the first night that Janice had gone home to get some rest. Two days later, his casket was carried to the grave site a hundred yards up a hill in a steady downpour of rain by his grandsons Ricky, Eric, and Dan, his great-grandson Denny, his great-grandson-in-law Bob, and me, his son-in-law. Dan and I returned home the next day, he to his job on an oil rig and I to keep the home fires burning, the customers happy, and my teaching props ready for classes that would start again January 15. Janice's mother was sick now, so Janice stayed to care for her and for her aunt Chloe before flying home after being away for almost a month.

Upon encouragement by a few faculty members, I applied January 23, 1991, for the position of head of the department of biological sciences at NSU. In retrospect, it seems stupid that I ever should have entertained such thoughts in view of past unpleasant interactions with the NSU administration. I went through the motions of preparing and submitting a vita summarizing my training, career, publications, contributions, certifications, honors, research grants, memberships, and letters of recommendation. From many applicants, half a dozen people were invited by the search committee to visit the department and present a seminar. I should not have been surprised at the extent to which the applicants went in preparing entertaining visual aids for their presentations. But I was surprised to learn that I should have been preparing a similar presentation, and by now, there was no such time available. Consequently, at my seminar, I walked in with a handful of reprints from years of research and proceeded to talk extemporaneously without helpful visual aids until my time expired. All my research was related to agriculture that was not of much interest to a department now focused on aquatic and marine research and associated problems with fish and oysters. Also, I was no longer a vibrant young researcher anxiously climbing the mountain for recognition. The job went to Dr. Tom Soniat, a New Orleans native with a BS degree from Nicholls, graduate study in Norway, and PhD from Texas Tech University. Tom's focus was on oyster research in which he had already made some real contributions as a young scientist and had served as president of the National Shellfisheries Association. His record qualified him well for the job from which he voluntarily stepped down four years later out of frustration with the administration and a desire to spend more time on research.

Josh had lived with his grandparents for nearly six years. He had a curiosity about everything and could entertain himself for hours without being lonely. He was neither shy nor aggressively outgoing. For a couple of summers, he enjoyed chasing and catching insects with a bug net and asking what they were whenever he found a different one. On one occasion as he chased a four-foot brown rat snake into a hole on the bayou bank, he grabbed it by the tail attempting to pull it out for a better look when I persuaded him to let the poor critter go. He went to the field with me rarely on short trips to look at problems or pests of wheat, corn, soybeans, or sugar cane. He attended two basketball camps at Nicholls State where he won the trophy for best free throw shooting. He attended church sometimes with his MeeMaw and sometimes with his PawPaw where he was attentive and well-behaved. One morning, when I was scheduled to preach at the First Presbyterian Church and before I realized it, he had assumed one of the two seats on the podium behind the pulpit where he sat smiling at the congregation during prelude music until, seeing my frustration, Sister Alma Zimmerman lovingly took him by the hand to her pew. After church one Sunday, he and I dug for earthworms and baited his hook for his first fishing experience. He's been an avid fisherman ever since; and my assistant, Steve Hoak, was his first dependable and ready fishing buddy. After moving to Nashville with his parents, he continued to look forward to summer visits in Thibodaux, fishing in the bayou, paddling his pirogue, and shooting at alligators with a pellet gun, almost as much as we looked forward to having him.

Nancy met Danny Stephen while attending job training meetings in Birmingham, Alabama, in 1990. They dated for a year and were engaged for several months before their marriage in a candlelight service March 16, 1991, in the First Presbyterian Church of Thibodaux where I was then an active member. A thunderstorm with heavy rain made it difficult that evening for guests and participants to get from their cars to the front door of the church and also from the church to the reception that followed at the NSU ballroom. Josh, age five, was the ring bearer at his mother's wedding. Following the wedding, the new couple made their home in Nashville, Tennessee, where Danny had graduated from David Lipscomb University and lived for a number of years. Joshua and his mother had lived with us from his birth until now. The new honeymooners had hoped that Josh could stay with his grandparents through the summer. However, Janice needed eye surgery for cataract removal and a lens implant and was afraid she wouldn't be up to playing mama for a while after her operation. And so Nancy and Danny drove down in July to get Josh and take him to his new home in Nashville. He had lived with us continuously for nearly six years, and we felt like we had lost another child. Our attachment to Josh has always seemed strange and sometimes irrational to his parents. However, they never have experienced that kind of pain. They now have three healthy children, including Josh who left the nest to study and play basketball at Abilene Christian University. Danny sells life insurance and other investment instruments while Nancy is a good wife, mother, homemaker, housekeeper, church worker, charity worker, part-time interior decorator, and baby-sitter.

After being my own secretary for twenty-five years, it seemed that the consulting business had grown sufficiently to justify hiring a real one, one who could help prepare and take care of field maps, spray records, soil-test results, fertilizer recommendations, mailing lists to potential customers, and a variety of related activities that I wanted to do but couldn't find time for. Three people were interviewed, and the best candidate was hired. In July 1991, seventeen-year-old Stephanie Borne was hired as secretary and later also became laboratory technician to count and identify microscopic nematodes, to analyze samples of sugar cane from seed plots for the presence of ratoon stunting disease (RSD), and to sort and crunch numerical data of various kinds. Steph had been an exceptional student with good computer skills who came highly recommended from the Louisiana National Bank where she had worked during her senior year in high school on a work-study scholarship. She was proficient in the office, the laboratory, and even the field on occasion when extra help was needed weighing sugar cane from experimental field plots.

September 27, 1991, the *Daily Comet* of Lafourche Parish, Louisiana published the second and final part of a guest column entitled "Crop Insecticides Necessary," in which I summarized the history and evolution of effective and safe insecticide use on sugar cane in the state in an attempt to allay rising concerns over pesticide hazards from citizens living in subdivisions that increasingly encroached upon agricultural areas. I also addressed recent letters to the editor, one of which had stated that insecticides were going into drinking water and wondered how many people would die from insecticide poisoning, in spite of the fact that periodic water sampling by the State Board of Health, together with water treatments, suggested that no such hazards existed. Another lady wrote of a terrible skin rash her son had experienced from walking in tall grass behind their house several hundred feet from a cane field that had been sprayed, and although her son no longer suffered from the rash, she wondered "what toxins could be lingering" in spite of the toxic plants and insects that may have been present there. Others were concerned about fish kills of which some occurred annually, but due more often to low oxygen tension in water during hot summer than to pesticides. I concluded by quoting a member of the U.S. Congress who recently had said, "There's no way to satisfy those who don't want to be satisfied."

Later that year, I was invited to present an invitational paper at a formal conference on teaching as part of the annual meeting of the Entomological Society of America held in Reno, Nevada, in December 1991. In opening remarks, I stated that "during more than thirty years of college teaching, I don't think that I ever have followed for two consecutive years exactly the same course outline. This might be proof that I haven't yet found the best way to do it or that I haven't got sense enough to know when I have found it. In either case, these are the things that I'm doing now, nothing really new, just the meat and potatoes of an old-fashioned introductory or survey course in entomology, similar to what many of you had once upon a time." It was interesting and nice to see some faces I had not seen since graduate student days. Also entertaining was my attendance at an unusual session on the subject of "menus for eating insects." Roasted crickets, grasshoppers, and fried caterpillars were passed about to be sampled by the audience. I sampled everything and concluded that, in a pinch, some of these might compete with crayfish.

Daniel had graduated from E. D. White Catholic High School in 1988, completed army basic training as an honor graduate in February 1989, and advanced individual training as a Lance Missile crew member a few months later before going to Germany, where he was stationed until his early discharge in June 1990, after nineteen months of service. He enrolled at Nicholls State for the fall semester but decided after several weeks that life elsewhere should have more to offer. He then worked offshore for several months on oil-drilling rigs, as a seaman on a supply boat in the Gulf, and as a truck driver delivering building supplies for Frost Lumber Company, before coming back in 1992 to work for his daddy.

University Retirement. Having determined to retire from teaching at the end of the spring semester of 1994, and knowing that I would teach my final entomology class in the fall of 1993, I had urged Dan to be sure to include my class in his schedule for that semester. I was certain that I taught the most practical aspects of the subject in a way that no one else for miles around could or would do and that this would be his best and probably last chance to have this opportunity. For what reason, I don't know; he didn't register for my class but took it the following year with a biology professor who had little interest in agriculture, knew little about the applied aspects of the science, and who was not an entomologist. Following that year, the university ceased to offer any course in entomology. I tried repeatedly, but unsuccessfully, to convince Dan of his need for a college education and of the fact that a future in agricultural consulting would be limited without it. Nevertheless, he believed that much of what I did was physical, and that the rest could be learned without need of a college degree, in spite of the fact that state laws now required a BS in agriculture or a related field as a prerequisite for taking mandatory consultant exams. As chairman of the LACA's legislative committee, I had been influential in bringing about these changes in the law. Whatever his thoughts were about such things, they didn't seem to occupy much of his time. He had a capacity for enjoying life without worry, which I had never experienced and could not understand. I loved him but couldn't reach him. We were living in the same house, but in two different worlds.

I resigned from Nicholls State University at the end of the spring semester of 1994 after thirty-eight years of faculty service, including eight years at LSU, to devote full time to my consulting business that had begun in 1965. It had been customary for retirement parties to be held in the home of some member of the department faculty. I requested that it be held toward the end of a working day in one of the biology laboratories on campus. I thought it was appropriately simple, and I didn't want anymore. They gave me a nice set of luggage that has served me well ever since.

My sixty-sixth birthday came and went September 20, 1994; and although I had a few aches and pains, I didn't feel like sixty-six. Right or wrong, I believed I was as good as most men at forty-six or certainly fifty-six. I even boasted to my customers that I would never retire from consulting, and that one day, a few decades hence, they would find me lying facedown somewhere on an ant nest in a cane field. I was still actively scouting fields, bouncing over rough terrain on an ATV or in a pickup truck, visiting farmers, planting and digging wireworm traps, cutting and collecting cane for RSD tests, exercising aerobically almost daily, doing yard work, and cleaning the swimming pool. I fully expected to be active for another decade or so and wondered why in the world I had waited so long to retire from the university.

The sugar cane disease of major concern was RSD, caused by a pathogen that occurs in the water-conducting vessels of the plant and that slowly increases in numbers inside the plant during the growing season. Plant pathologists had finally developed a fairly sensitive chemical method for detecting the presence of the pathogen in cane juice. There was much talk about the effectiveness of the technique, but no one was offering a service to farmers to help them identify healthy seed cane for planting. So I read up on the subject, and Stephanie and I visited with USDA pathologists to learn how to do the testing. After a while, we began to feel as competent as anyone in using the technique and reading test results.

In August 1994, I equipped a laboratory and began offering this disease-testing service to customers. We began to find RSD here and there in places that surprised some farmers, pathologists, and, even on occasion the, producers and vendors of disease-free seed cane. Some of the pathologists and vendors naturally supposed that we were reporting erroneous results and that we weren't capable of doing this work. However, when we compared our results, from divided test samples, with those of the pathologists, there was little difference. We all had problems deciding whether the colors on the test papers at very low pathogen levels should be judged positive or negative; and so results were not always clear-cut. Nevertheless, the occurrence of definitely positive results seemed to justify continuing the service, which we did for another three years.

These events spurred interest in the industry in setting up a laboratory at LSU to provide a free service to growers beginning in 1997. As a result, I didn't promote my RSD service after that year but continued it on request for those who wanted it. Since that time, pathologists have concluded that the method is not generally sensitive enough to reliably detect the disease until after the planting season is over. And it is now recommended that testing be done during harvest, after planting is finished, in old fields being harvested for the third time or more. Therefore, it now is used mainly as a quality control method to assure that RSD is not increasing on the farm. I sometimes wonder if there would be such a service available to growers today if I had not made the initial leap. Of course, from a business standpoint, my decision was no better, although less expensive, than my earlier venture into household, structural, and horticultural pest control.

My greatest physical concern at this time was an exaggerated curvature of my penis during erection caused by Peyronie's disease, the symptoms of which, according to a secretary in Arkansas, also afflicted President Bill Clinton. At any rate, it made my wife uncomfortable, and I was determined to do something about it. Wednesday, January 4, 1995, I lay naked on a table at Ochsner Clinic for a presurgical examination and demonstration by Dr. Harold Fuselier (head of Urology at Ochsner Foundation Hospital) for the benefit of resident interns. As I was wheeled into the room, the first images I saw were the faces of at least a dozen male and female interns smiling down from the observation gallery several feet above my head. The doctor explained that Peyronie's was not really a disease but an abnormal condition in the normally spongy erectile tissue of the "corpora cavernosa," caused by development of scar tissue there, which caused the penis to inflate abnormally like a balloon with a patch on one side. By artificially inducing an erection, he demonstrated that blood flow to the organ was very strong and more than adequate. As he held my organ with one hand and pointed with the other, he explained how tomorrow he would break up the plaque (scar tissue) before inserting an inflatable prosthesis into the "corpora" of my penis, attach a saline reservoir to the lower "rectus" muscle of the abdomen, and connect all of these together with a syringe-type pump implanted in my scrotum. Surgery was performed the next morning, and I was released to go home two days later. After two-and-a-half weeks of convalescence, I was on the road again and happy with the thought that my wife would no longer suffer pain due to my crooked organ and also that I was now potentially an all-night-long or sixty-minute-man, whatever the situation might require.

On the morning of February 14, Valentine's Day, 1995, Janice found a note on the refrigerator, which sent her to a chair in the living room, where another note sent her to the piano in the library and music room, on which a note was found that sent her to the clock on the mantle over the fireplace in which she found a pearl cluster ring. She appreciated the gift and the method of presentation and gave me in return a beautiful shirt and card saying, among other things, "I love our life together."

For the first time since childhood days, my brothers and I all got together for a weekend to renew acquaintances. I left Thibodaux Friday morning, April 28, 1995, in my pickup truck headed for the Joe Wheeler State Park in Rogersville, Alabama, arriving in late afternoon. Needham had already dined and located our room. He accompanied me to the lodge restaurant where I feasted on blackened catfish before Rob arrived about 9:00 PM from the emergency room where he had gone with a surprise gallbladder attack. We all three had been achievers in our professions; but only Needham had largely escaped, by comparison, some of the unhappy consequences of our early childhoods upon marriage, family, and professional life, due to his early experiences with panic attacks and psychoanalysis.

We visited, recalling childhood memories, and also listened to Ellen Kreidman's tapes, entitled *Light Her Fire*, until after midnight when Rob returned to his nearby home and wife. Needham and I talked into the early morning hours. The next morning, we drove to a Waffle House for breakfast before returning to the lodge where Rob joined us later and we spent most of that day and evening again visiting and listening to the

Kreidman's tapes. Rob left for a while to see patients and returned to visit a while more before going home. After checking out of the lodge Sunday morning, we went to Rob's house and visited with our sister-in-law, Barbara ("Babs"), and nephew, Bryan. Babs, a very talented artist, insisted that we take at least a painting each for the past Christmas. We each took one and bought another.

I departed after lunch for Huntsville, Alabama, to visit Mother in her "assisted living" apartment. We discussed my brother Needham's book, "Panic: An Odyssey—Anger in Disguise" (published only after her death by Dorrance Publishing Co., Pittsburg, PA, 2005) in which she was offended deeply by his memories of a painful childhood. I tried to help her understand and feel better about it. She wanted me to stay and sleep, but I refused to take her bed and declined to sleep on the couch in favor of a nearby motel. However, we had breakfast together in her apartment the next morning before I made the long drive back to Thibodaux.

During 1995, Dan, Steve, and I did all the fieldwork; and Stephanie and I ran the office and laboratory. The collection of soil samples from the fields, shipping them to the A&L Laboratory, developing recommendations from results and presenting these to farmer clients occupied most of our time during the winter months. In late winter and spring, I assisted some clients with their concerns about weed control. As soon after this as possible, we began to visit clients and prepare field maps for the approaching field-scouting season. I was now on medication for glaucoma in both eyes and, in April, started seeing a chiropractor for back, hip, and leg pain, but without much relief. Then in June, scouting began in earnest when Dan, Steve, and I each had a route of farms to cover daily. Our general procedure was to rotate routes from week to week so that during a three-week period, each of us would see all farms and fields. In this way, I could keep up firsthand with circumstances and events on all our farms.

The 1995 season was Steve Hoak's tenth year in which he had worked each year for roughly six months with me and six months in Florida with Glades Crop Care Inc. Steve had become a seasoned scout and fieldman—much liked, well-respected, and trusted by all clients. He was tired of moving back and forth each year like a migrant worker. I valued his presence highly, got along with him well, and the feeling was mutual. However, there was much concern at this time about the future of the Louisiana sugar industry and whether it could survive after NAFTA and in the face of increasing competition with cheap sugar from third world countries. We discussed these things a number of times, and Steve indicated that this would be a year of decision for him as Glades wanted him badly to come with them full-time. I didn't want to lose Steve, but even as much as our business had grown since my retirement from teaching, it would have been a serious strain to pay him on a year-round basis, particularly since the business was now supporting both Dan and Stephanie year-round. So when the scouting season was finished, we had another conference at which I offered him $800/week for five months or $30,000/year for 1996. Glades could beat my offer and also offered health and retirement benefits that I could not match. On Friday, October 13, after visiting one customer, he drove back to the office and resigned. He stated that his main reason for doing so was the uncertainty about the future of the sugar industry. However, he also had told me earlier that he would never work with Dan in my absence and that it was obvious to everyone that Dan got away with behavior that would have gotten any other employee fired in a minute. Unfortunately, I could not dispute his statement. Janice and I later attended his wedding in Florida, and we still communicate with each other from time to time.

Total expenses were unusually high in 1995. Home repairs amounted to thousands of dollars above budget for such things as roof, fence, and door repairs, replacement of ceramic tile in bathrooms, window replacements, and termite repairs in addition to new furniture and skylights for the upstairs TV room, California Closet Company shelves and hanger bars in the master bedroom closets, and refurnishing and moving the office from the second floor to the downstairs new room. Two microscopes plus other laboratory equipment alone had cost nearly five grand. Our grandson, Denny, went to Salt Lake City to begin two years as a Mormon missionary for which we were obligated to help support. This also had been a year of frequent travel during which we had driven three times to Delhi, Louisiana, including a Thanksgiving visit with Jan and her family, and four times to Nashville, including a Christmas visit and others to help with Nancy's new baby, Jake. When, in reviewing cancelled checks, I discovered that we had spent over a thousand dollars decorating the house for Christmas, I was appalled and simply exploded. Afterward, I suggested to Janice that it's not good for either of us to sit around sulking and, if she couldn't be happy here, she might consider spending Christmas with her mother. Two days later, I bought her a ceramic church as a peace offering to start her "snow village" collection, and we spent Christmas together.

After dating and partying for several years, Dan finally met Christina Adams in 1990, and she became his "one and only." They were married at Saint Genevieve Catholic Church in Thibodaux at 7:00 PM Friday, January 5, 1996, with a reception following at the NSU ballroom. The rehearsal supper on the previous evening at the Sheraton was especially memorable for the numerous anecdotes and testimonies of friendship offered in sequence by a large number of their friends.

Two years after retirement from Nicholls State, my consulting business had almost doubled, and we were scouting more than fifty thousand acres of cropland weekly with three and sometimes four trucks and ATVs running daily in the fields, soil testing nearly ten thousand acres annually, and starting to test seed cane for the presence of RSD. For 1996, I was offering a choice of crop management packages combining any of our traditional services with other options, including RSD testing. Facing this scouting season with a bad back and a relatively inexperienced team of field scouts, I felt more challenged than ever before. Although this would be Dan's fifth season of insect scouting, the two new employees had no previous experience. Dan had learned a lot, but it takes a lot of time and accumulated circumstances for a farmer to trust someone else's judgment regarding the treatment of his crop. Although I was dealing with more back and leg pain each month, I had no idea how bad matters would soon become.

At the February 13, 1996, workshop meeting of our state consultants association (LACA), the association voted to send me to Washington DC, to accompany State Commissioner of Agriculture, Bob Odom, and cotton consultant, Ray Young, to convince the EPA of the need for more rather than fewer insecticide tools to work with. I was up at 5:30 AM Monday, February 26, to catch an 8:15 AM flight to Washington with Ray Young and Bobby Simoneaux, Commissioner Bob Odom's representative. We attended a 1:00 PM meeting, which lasted almost an hour, with EPA pesticide registration officials. Most of the time was devoted to registrations on cotton, after which I was permitted to make a statement about sugar cane. I made a plea for new registrations of the much safer insecticides, Karate and Confirm, and for more rather than fewer tools to preserve our sugar cane IPM system in view of the fact that so many restrictions now had been placed on the use of Guthion that it was practically useless in most situations. My statement was favorably received with indications that our desires would be forthcoming.

Back Surgery. Wednesday, February 21, 1996, Dr. Miceli, an orthopedic surgeon at Ochsner Clinic, told me that x-rays and an MRI showed a slightly herniated vertebral disc and other signs of deterioration in my lower back and referred me to appointment with a neurosurgeon. Two days later, Dr. Voorhies, head of Neurosurgery, diagnosed my fourth lumbar vertebra as being slightly out of line but didn't believe that surgery was indicated at the time. He sent me to physical therapy for a list of exercises to be performed daily.

Dan spent most of March studying for his state agricultural consultant's exams that were needed to qualify him to sign recommendations. In early April, I started classroom training sessions, of which there were several, for our two new employees, Jason Pontiff and Louis Guillot, before taking them to the fields to become familiar with farms and their locations two months before the regular scouting season would begin. Dan said that I was wasting time showing them slides and lecturing and didn't believe they needed two months to learn where all the farms and fields were. I disagreed with him more or less continuously about this and other things as well and sometimes was forced to do so within sight and hearing of the other employees.

Back and leg pain was becoming more intense with each passing week. By late June, I was switching back and forth between aspirin, Aleve, and Voltaren looking for more relief but finding little. Some days, I was in the field more than thirteen hours. My ninety-six-year-old mother was not doing well and was becoming increasingly depressed in her assisted living apartment in Huntsville, Alabama. I began calling her regularly in April trying to cheer her up. On one occasion, she said that she had been enjoying painting but felt guilty for enjoying it.

In late April, Janice left to visit Nancy and family in Nashville and then on to Raleigh to visit her mother and family there before returning three weeks later. She and I then drove to north

Louisiana to Jan's home. And on Sunday, we—with Holly, Heather, and Heidi—followed in our car behind Bob, Jan, Amanda, and John Henry to Clinton, Mississippi, and the church stake center there to attend stake conference. Our granddaughter, Holly, gave her testimony at the conference with poise and sincerity, and we witnessed as she was set apart for an eighteen-month church mission to Ogden, Utah. We returned to Jan's place for dinner before heading back to south Louisiana and home by late evening. Tomorrow would be another day in the field with Dan and Jason going in one direction and Louis and I in another.

In early June, I spent an entire Saturday attending a tutor-training workshop, sponsored by the Lafourche Council on Aging, to learn about methods for teaching the illiterate to read, and volunteered my services as a tutor to begin in September, not realizing how many other matters would be occupying my time come September. Janice and I had seafood dinner that evening at Politz's Restaurant. For Father's Day, Nancy gave me a book, entitled "My Aching Back," which I read with appreciation and interest, but without much relief.

Regular crop scouting had started in early June, and we were making weekly rounds without serious problems until Friday, July 12. As I drove through a thunderstorm from field scouting in St. Mary Parish and listening for my employees to check back and forth on their radios, I was startled by the voice of a farmer client, John Burt, saying, "Dr. Long, are you there? There's been an accident on my farm. Jason was hit by a truck in the field. But he'll be all right. The ambulance has taken him to the hospital." Jason had suffered two broken bones in his left leg below the knee, a cracked rib, and numerous cuts and bruises. The collision of his ATV at a field road intersection with the farmer's son's pickup truck, which had been traveling much too fast, had tossed Jason and the ATV about forty feet from the point of impact and deposited him on top of a fire ant nest. The ambulance had arrived in a few minutes by which time he also had been stung severely. I was relieved and happy to find him at the hospital in surprisingly good spirits and very sorry to lose his help for the rest of the season. Jason Pontiff had been an outstanding high school athlete and was a student studying at Nicholls State to be a football coach. A few days later, he came to the office on crutches with his initial medical bills, which my workman's compensation insurance covered. He recovered completely within several months and with no desire to pin liability on anyone for damages or pain.

Janice left again in late July for visits in Nashville and Raleigh before returning home in early August to find catch basins and dirt fill in the front ditch by the highway and a semicircular concrete drive being poured in front of our home for appearances and also to better accommodate traffic to and from the house and office. She, with Dan and Christina, left again at the end of the month for a week on the beach at Gulf Shores, Alabama. I finally got away in midafternoon arriving there in time to go with the family to BJ's Restaurant where Danny treated us all to dinner. I appreciated the party, but penny-pincher that I am, the thought occurred that a can of tuna with fruit salad would have been healthier and less costly than the hundred dollars spent. Monday morning, I returned to Thibodaux to scout soybeans and recommend spraying for stinkbugs at Edgard and Vacherie. My back pain was so intense now that I often had to stretch out on the ground for relief, regardless of circumstances. Janice et al. returned from the beach Saturday, September 7. Four days later, I forgot our forty-third anniversary until it was almost over. I wrote my wife a love letter and spent the rest of the evening trying to make up for my forgetfulness.

With most of the field scouting of sugar cane completed now, except for a few hundred acres of soybeans, I hastened to see my primary physician at Ochsner Clinic. On September 16, 1996, Dr. St. John sent me for the earliest possible appointment with

Neurosurgery that would be September 27. Meanwhile, I called Mother, her doctor, and her nurse to learn that her situation was not good and that she would be in the hospital for at least several more days. Thursday, September 19, I was up at 4:30 AM to pack and leave for Huntsville, Alabama, arriving in late afternoon to find Needham and Winnie with Mother in her hospital room. I'm sure she recognized me at the moment and knew what was going on a great deal of the time, but it was impossible to understand her as she tried to talk with a stomach tube in her nose and throat. Nancy came from Nashville with Danny and Josh to visit for a while. I alternated between holding Mother's hand and doing stretching exercises on the floor to relieve my own pain before going to her Wyndham Park assisted living apartment for rest.

Needham, Winnie, and I spent the day Friday at the hospital to learn that Mother would not recover from the effects of pneumonia and diverticulitis of the colon that had shut down all intestinal function. Rob and Babs arrived in the afternoon. Mother's "living will" forbade artificially prolonging her life under such circumstances; and we, her sons, advised the doctor to discontinue antibiotics and glucose. Rob and Needham removed her stomach tube at which she showed signs of relief. She was conscious off and on during the afternoon and evening, but I believe she heard and understood a great deal even when appearing unresponsive. At one point, she raised her hand as if signaling her desire to be included in the conversation. There were numerous visits from friends during the day and early evening.

Everyone was exhausted, and I was in severe pain when we finally employed a sitter to spend the night during which Mother was mostly unconscious. Saturday, there were again numerous visitors while Mother remained comatose all day. Janice arrived in our van from Thibodaux. After the sitter came at 7:00 PM, we three sons and our wives went to dinner together before assembling in Mother's apartment to discuss disposition of belongings, furniture, and cleanup. At almost midnight, I was awakened by a call from Mrs. Edwards, the sitter, to hear that Mother's breathing had become very erratic. I felt guilty for not getting up immediately but rationalized that since she wouldn't know whether I was there or not, I would be foolish to suffer the pain that rising and walking would cause. Two hours later, at 1:30 AM Sunday, September 22, 1996, Mrs. Edwards called to say that Mother had gone very peacefully. I dressed and went to the hospital where we paid the sitter and picked up Mother's things. I kissed her goodbye while she was still warm, thankful that I had held her hand during the last three days and sorry that I was not holding it when she left.

Brothers and wives spent the day packing boxes, dumping trash, and sorting papers, letters, and pictures. Nancy came with a borrowed truck to take Mother's china cabinet to Nashville. The three couples dined at Landry's restaurant before retiring. Janice was up early Monday morning to leave for Raleigh to be with her mother during surgery on Tuesday. Rob and Needham both were so concerned about my problem that they arranged for me to see Dr. Frank Haws, a Huntsville neurosurgeon, that morning. After examining me and taking several x-rays, Dr. Haws recommended that I get an MRI that he was certain would indicate the need for immediate surgery. Rob successfully negotiated getting my Ochsner-65 HMO to authorize an MRI in Huntsville where it was done Monday evening at 9:30 PM. Nancy arrived again from Nashville and Dan and Christina from Thibodaux for visitation at the funeral home that evening as other family members also began to assemble.

Tuesday morning, September 24, I bathed and dressed at Mother's apartment and drove twenty miles to Decatur for breakfast at the Holiday Inn with Nancy, Dan, and Christina. We attended the funeral at the funeral home and cemetery before attending a luncheon at the Pryor home with the rest of the family and friends. I picked up my MRI film at Bioimaging and went back to Mother's apartment for goodbyes to family who were loading cars and a trailer. Dan, Christina, and I departed for Thibodaux where we arrived two hours after midnight as I lay on the backseat of their car. I worked at my desk all day Wednesday while Steph worked at the microscope and Dan scouted soybeans. A call to Janice indicated that Mother Rogers was doing fine after surgery.

Thursday, September 26, I was up at 5:30 AM for Dan to take me to Ochsner hospital for a myelogram and CAT scan imaging of my back. Because of a bad headache caused by leakage of spinal fluid from a puncture by the myelogram needle, I took a room for the evening at the nearby Brent House. Dr. Rand Voorhies told me the next morning that I had a number of back abnormalities that would require major surgery with fusion of the lower vertebrae to correct them all. Since most of my back pain was really in my buttocks and associated with numbness and loss of strength in my legs, he recommended that we first try the less drastic option of removing the causes of pressure on nerves and the spinal canal between the L4 and L5 vertebrae, particularly on the left side. I agreed, and surgery was scheduled for October 7. Stephanie and Dan came to take me home.

During the next few days, I was on my back most of the time while unsuccessful attempts were made by my brother, Rob, to get the surgery done sooner. Meanwhile, I called our complete-service farmer customers to inform them of my approaching surgery and to assure them that their needs would be taken care of. By calling Janice, I learned that Mother Rogers was recovering beautifully and would have her surgical clamps removed that day. Monday, September 30, Dan and Christina brought a pizza after work, and we watched a football game on TV. The following day, I stayed on my back as much as possible and heated a can of chicken and dumplings in the microwave oven for supper. Dr. Connelly, another neurosurgeon, called on Wednesday to offer surgery on Friday, but I elected to keep my Monday schedule with Dr. Voorhies. Apparently, they had already made the change, thinking that I would accept Connelly's offer, which I did not want to do. After considerable confusion, it was finally agreed that Dr. Voorhies would operate on me Monday, as originally planned, but that I would be his fourth surgery of the day even if it took until midnight to complete.

Thursday, October 3, I lay on the floor most of the day but made three trips to town for errands and groceries. Nancy called her mother in North Carolina to tell her that she ought to be home with her husband for his surgery. This upset Janice who felt that she was committed to help her mom recover from recent surgery. My preop activities began Friday with an EKG, blood test, meeting with anesthesiologist, signing of papers, and talk with Dr. Voorhies. Saturday, I lay around the house and made some phone calls while Dan and Christina tended to chores. Sunday, I wrote thank-you notes for flowers and pallbearers at Mother's funeral, called some customers to reassure them about my situation, and called Walter Hackney, my pastor, who prayed for me over the phone and promised to see me tomorrow to administer to me according to James 5:14-15. I also called my brothers, my wife, and Nancy and Jan. Then two of my Mormon friends, Lawrence Bergeron and Boyd Atterbury, called wanting to come and give me a healing blessing. I thanked them and told them that Rev. Hackney was going to do that for me tomorrow, but they insisted that they would like to do it tonight also. And so they did.

Monday, October 7, I called Neurosurgery at Ochsner and left a message for Dr. Voorhies to the effect that my symptoms had changed, that I was having pain and numbness on the right side as well as the left, and that I would like for him to enlarge the surgery to include decompressing the L5 nerve roots on both left and right sides. Dan drove me to the hospital in the afternoon where I was dressed for surgery when the operation was cancelled at the last minute due to one degree of body temperature. They took more lung x-rays and blood and urine samples before sending me home with a prescription for yeast in my stools. I had been having a problem with diarrhea for about three weeks. The next day, Dr. Wilson called to say that blood and urine tests were all good and I was being turned over to Dr. Voorhies once more. Then Dr. Voorhies's nurse called for me to come in October 15 to see the doctor again, which meant that surgery was still more than a week away. Dan brought crutches, which made matters a little easier, and did some grocery shopping for me.

I was up at 5:00 AM, Tuesday, October 15, for bath, shave, breakfast, and scripture reading before calling Janice, who was on her way home, to tell her I loved her and to be careful. At Ochsner Clinic, I hobbled on crutches from my 10:30 AM appointment with an internist to a colon and rectal exam at 1:00 PM and finally a visit with Dr. Voorhies at 4:00 PM. We discussed the surgical game plan, and he told me that again I would be scheduled for his fourth surgery on Thursday. Janice was home when Dan and I returned from New Orleans. Wednesday, I wrote and faxed a letter to Dr. Voorhies and received one from him in return outlining the game plan agreed upon the day before. While Dan visited farmers, Steph worked at the computer, and Janice went for a hairdo, I lay on the floor and listened to news. In the evening, Janice fixed dinner, and we watched the presidential debate.

Thursday, October 17, the long-awaited day finally had arrived, and I was at the hospital before noon. I was sent to surgery at 3:30 PM and to recovery three hours later, where the smiling nurse looked down and said, "You were asking for Janice while regaining consciousness." Before midnight, I had walked to the bathroom with Janice's help and in the hall with her holding me up by one arm while I hung on to a handrail with the other. I could already tell that leg pain from nerve pressure was gone. Dr. Voorhies visited the next morning to describe the operation. Some things had been encountered that were not anticipated. He had done laminectomies on both sides at two levels, L4 and L5. He had removed bony spurs and synovial cysts from both sides and trimmed a bulging disc between S1 and L5. He was glad that he had worked on both sides and felt that he had accomplished a lot of nerve decompression. He sent me home to rest and to start walking a little at first, increasing each day as much as I comfortably could do. He later told me to try to walk at least two or three miles daily for the rest of my life. After a week, I could walk a quarter of a mile without pain but with some weakness and occasional numbness. By late November, I could walk over two miles. Janice helped me to bathe daily and waited on me like a full-time nurse.

Dr. Voorhies saw me again in mid-November to offer some general advice and to say that I wouldn't need to see him again unless problems arose. I had begun again to experience pain on some days, but not always. One day in early December, I was afraid to walk due to pain. Dr. Voorhies suggested leaving off the exercises for several days and walking only. Then in mid-December, he changed my medication from Lodine to six days of cortisone to see if that might relieve some of the numbness I was experiencing. Finally, we concluded that my lower back and other associated parts might never be normal again but that I should continue to walk and exercise with discretion, unless or until pain, weakness, or numbness should force me to return for more radical surgery.

Calvin Viator came by to get an outline map of Georgia Plantation as, due to my incapacity, Bryan Harang had asked him to estimate some crop damage from accidental herbicide drift from a nearby farm. Dan was upset that Bryan had asked Calvin rather than him to do this and I could not explain, to his satisfaction, why I was not perturbed or surprised about it. I walked two miles in the driveway as Janice attended a celebrity brunch at Laura Badeaux's house where she was joined at her table by Eric Paulsen of WWL-TV. She enjoyed this very much and described the occasion to me with considerable animation. I listened with a show of interest but was privately grateful that I hadn't had to be there.

Dan called on Monday to say that, if he wasn't needed badly, he had other things to do and would visit farmers tomorrow. I told him that visiting farmers was not on his schedule for the week but that I would see him tomorrow. He arrived the next morning in a bad mood, resentful, and argumentative. We had a lengthy verbal exchange during which he said that I didn't give him enough decision-making authority, that I didn't do enough to teach him what he needs to know, and that I did nothing to promote his reputation in the customers' eyes. I explained that he would have to earn the customer's trust and respect without much help from others, that it would require more years of harder work and less play, and that the more time he could spend with me, the faster he would learn. I said that I was tired of arguing every day about every thing and that he should quit now and find another job if he couldn't be happy with the fact that I would continue to review decisions made by him and all other employees as long as I had anything to do with the business.

For the next several months, Dan was more or less occupied with soil sampling, tending to vehicles, and visiting growers. During the harvest season, a few customers expressed concerns about some real or imagined crop damage, the thoroughness of our scouting by all the new personnel, and their unhappiness with seeing different and unfamiliar faces in their fields each week. We visited with some of them to allay their concerns about crop damage. This led me to consider that in the future, each scout could check the same farms each week and that I could follow making timely appearances as often as possible on farms where my presence was most needed or appreciated.

Before the year had ended, Stephanie advised me that she would resign next April to attend LSU for studies in medical technology. She would be happy to help train a new secretary/technician. And so after almost six years, I would be losing Stephanie. Jason advised me that he would be willing to work again next year from June through July but would have to quit in early August for football. Louis would be able to work if or when we were in a bind, but he had other competing activities that he would like to pursue. Realizing that I would never fully recover from back problems, and facing a new year with the need to find and train inexperienced scouts and a new secretary, I was discouraged and a little depressed.

Year of Decision. In January 1997, Janice accompanied me to a five-day meeting of the National Alliance of Independent Crop Consultants in San Antonio, Texas, and in February to three days and two nights at the Grand Casino and Hotel in Biloxi, Mississippi, as guests of the Valent Corporation at their consultants' seminar. The ladies all had great times. While Janice and I attended the Biloxi seminar, Dan and Christina were attending another consultants' seminar and skiing at Lake Tahoe.

Stephanie listened in late January while I interviewed Judy and Melissa as her potential replacement. Judy Price was a mature, refined mother of two and wife of the pastor of the local Methodist Church. Melissa was a single, sexy, hot-looking babe, who had an apartment near the university. Melissa's short skirt and low-cut blouse aroused some basic instincts, but Judy was clearly the right choice. Stephanie laughingly said, "Dr. Long, you need Judy; Melissa is not what you need." Of course, I already knew that but was amused by her concern that I should make the right choice. Judy started training with Stephanie February 4, 1997. We didn't know at the time that some of the things Stephanie was teaching her, such as lab procedures for RSD testing and record keeping for farmers, would be reduced or discontinued as unprofitable or impractical activities in the very near future.

For Valentine's Day, I suggested that we visit Nancy et al. in Nashville, but unknown to Janice, I had made reservations for two nights at the Opryland Hotel. Since she had previously enjoyed visiting the hotel grounds several times and always wanted to see it again, I suggested a sight-seeing tour there before going to Nancy's house. When I drove under the porch and gave the valet my keys, she was quite surprised. By the time we left the Grand Ole Opry the following evening, we had seen the Opryland Hotel grounds, restaurants, shops, entertainers, and enjoyed another romantic weekend.

In March, we drove to West Monroe, Louisiana, for supper with Jan and her family before riding with them the next day to the Jackson, Mississippi, airport to meet Denny returning from his church mission in Canada. We arrived in time for a 3:00 PM flight, but Denny arrived at 9:17 PM due to a missed connection in Cincinnati. While waiting, we took the family to the Western Sizzlin restaurant for dinner, during which Bob's hopes for Denny to attend Louisiana Tech on a football scholarship were mentioned. The coach knew of Bob's football prowess in high school and had told him that he would take a look at his son. Denny didn't really like football and had not excelled as a high school athlete. I thought Denny would be smarter to earn college access via the military and the GI Bill. Next morning after breakfast at Jan's place, Bob took Denny to run some errands, and they were gone for the rest of the morning. So we didn't get to visit much with Denny before departing. However, in April, Denny and I drove a truck to Florence, Alabama, to pick up furniture from

Mother's apartment that my brother Rob had stored in his garage to hold for Jan and her children. Denny and I had ample opportunity to visit during this trip. He later earned the GI Bill, graduated from Louisiana Tech University in computer science, and took a good-paying job as a computer engineer in Houston, Texas.

For the next three years (1997-99), we held on to our customers and maintained a service area of over fifty thousands acres of growing crops. Confirm was a new highly effective insecticide and more environmentally safe than any we had used before. I had tried it in my own test plots before farmers or other consultants knew much about it and had determined that it would give excellent control at half the recommended rate. The chemical was priced to be competitive at an eight-ounce rate of application. I would recommend it at one-half that rate with equally good results. Other consultants and farmers were tentative about using it much at all until they had seen more of it. I had also determined to my satisfaction that a chemical wetting agent, different from and cheaper than the one recommended by the Rohm & Haas Company manufacturers, worked very well. The company representatives invited sugar cane consultants to a meeting April 23, 1997, to discuss these matters. I shared my information with the group but was the only consultant who left the meeting determined to follow my own thinking on the matter.

For the next three summers, our clients benefited considerably from dollars saved, in comparison to other farmers, by using the four-ounce rate of Confirm when spraying fields for SCB control. To combat my success with the low rate of Confirm, a competing consultant reduced his recommendations from eight to six ounces, and the word began to spread among aerial applicators and farmers that the lower four-ounce rate didn't last as long and resulted in the need for more spraying. Our records showed that this was not true, and by sticking to our guns, we continued to have a lot of happy customers in this matter. At the same time, we were again going against LSU recommendations. Chemical distributors were unhappy that they were selling less Confirm to our customers than to others. The Rohm & Haas Company, which owed us a vote of thanks to begin with, were now also unhappy over our success in recommending the low four-ounce-per-acre rate. The situation gave new meaning to the old saying, "What's good for the goose ain't necessarily good for the gander."

Steph worked her final day on Friday, May 4, 1997, after nearly six years with the company. Two days later on a Sunday afternoon, we had a going-away party for her in our backyard. Dan came at noon with Jason and a new employee, Randy, to boil crayfish, and Janice made her best crayfish dip. I set up horseshoe stakes and tied balloons to the mailbox. Other guests, who attended, in addition to the guest of honor, were her parents, Kerry and Maudi; her boyfriend Mike Gautreaux; her sister Stacy and boyfriend Cecil; Jason's girlfriend Monique; Dan's Christina; and my new secretary Judy and husband Wybra Price. I presented Steph with a gift certificate and tried to make a speech about her contributions to the business but got choked with emotion before saying anything significant. We talked, played horseshoes, and ate crayfish for three hours before Stephanie was gone. The party had gone pretty well, I thought, in spite of my foibles. We called Denny in the evening to wish him luck as he would leave for army basic training the next morning.

The recent hiring of Randy Richard, a classmate of Dan's and an ag major at Nicholls, was encouraging, but he had a lot to learn to be of real help and not much time in which to do it. In mid-May, I took Randy and Jason for a day of training to the USDA research laboratory in nearby Houma. This was a quick review for Jason and a first-time opportunity for Randy to see what a SCB looked like in all stages of its development. The laboratory personnel were happy to make several presentations for our benefit, after which we went to the field to dissect SCB-damaged cane plants and observe the pest in its natural setting. The next day, Randy and I scouted corn at Edgard and visited farmers on the west bank of the river while Dan worked with Judy on field maps. Work on map preparation had been in progress for weeks, but for the next ten days, there was a frenzy to finish the job before scouting began.

Sunday afternoon, at my pastor's request, I went to the church where Rev. Hackney introduced me to Elaine. Elaine was a poor retarded middle-aged woman with no teeth and ragged clothes and who lived in a shack with an abusive husband. Walter wanted me to help her learn to read, which I agreed to do. I started meeting her at the church on Sunday afternoons for study sessions with assignments. Later, we moved our meeting place to the public library where we met twice weekly for a while. Elaine was enthusiastic at first, but after several months, her enthusiasm began to wane and she started missing appointments, occasionally at first and more frequently later. The first time she missed two whole weeks she was in a battered-women's shelter. Afterward, she was often with a boyfriend or looking for one. Finally, after approximately a year, during which she had made little if any progress, she gave it up. I felt badly for having failed her, but I wondered if anyone else could have done better. Janice had been concerned about my being somewhere with Elaine two or three hours a week until I pointed her out one day walking along the street.

Josh, now almost twelve years old, chose fishing in Louisiana for the summer rather than baseball in Nashville. Janice met him at the New Orleans airport June 23, and the next day, he and I fished at Point-au-Chenes where he caught three trout that were used for instruction in the art of filleting. On another day at Point-au-Chenes, he caught a sixteen-inch flounder and some crabs that we boiled and ate that evening at home. He and I made a few bored-joint counts in sugar cane at Acadia Plantation and later played a little poker in the evening. On July 4, we had a family cookout in the yard with Dan and Christina, her parents Bobby and Rose Mary, Christina's sister Tammy and husband George, and Nancy and Danny, who came down from Nashville. Josh and his MeeMaw also enjoyed time together during his monthlong visit until she had to return him to the New Orleans airport July 19 for his flight home.

Field scouting began in earnest Monday, June 2, 1997. Dan and Jason went together in one direction and Randy and I in another. Where we went was now determined to some degree by customer demand. Some farmers told me plainly that as long as they knew I was in the neighborhood, they could trust the scouting. Others were satisfied with my assurance that Dan's experience was all that was needed for most situations and that if special problems arose, I was always available. In mid-June, I was a guest speaker at the Bayou Young Farmers Association dinner that was also attended by Dan, Jason, and Randy in their company uniforms. A week later, in front of Jason and Randy, Dan objected one morning to my directions for the day, resulting in an argument between the two of us. He wanted to give orders for the day and was unhappy about having to take them from me. I urged him to look for another job.

Jason helped Dan to cover more ground faster, especially when they traveled in separate vehicles. Randy learned rapidly to make dependable counts so that I could leave him on an ATV for a couple of hours while doing something else or just resting my back. When he called on the radio, I'd meet and check with him in some of his problem areas. Then we'd decide together what fields, if any, needed treatment before marking the map and reporting to the farmer. Jason had to quit, as he had warned, after July. This left us with two healthy scouts and one sixty-nine-year-old cripple trying to appear healthier than he was, with little more than a month to go before most scouting would end. Judy had handled the office, with little help from me, for record keeping, payroll, etc. Steph came back during a break from classes to help during a busy week in August with RSD testing. By early September, we had completed another season of cane scouting, Randy was off the payroll and back in school, and we had a few acres of soybeans to scout for another month after which Dan would start collecting soil samples.

During the summer, Janice traveled in June to Raleigh and Nashville and home again. Also in June, she and I had attended a three-day meeting of the American Society of Sugar Cane Technologists at Fort Walton Beach for a combination of business and pleasure. She spent the first week of September on the beach at Gulf Shores with Nancy and family. Four days later, I surprised her with a trip to Saint Francisville where we celebrated our forty-fourth anniversary at the Shade Tree bed and breakfast, a beautiful spot in the woods. We stayed at the Gardener's Cabin there and dined at Mattie's House restaurant on the grounds of the Cottage. A week later, we visited the Biloxi Grand Casino. The hotel was full, and we had no reservations but enjoyed a buffet dinner at Montana's restaurant and an evening at the Holiday Inn.

In early October 1997, Janice and I drove to north Louisiana to accompany Jan and Heather on a trip to the hospital. Heather, for several years, had needed surgery on her deviated nasal septum to improve her breathing. We had urged action in the matter and now went to help. She went to surgery Thursday, October 2, at 7:00 AM for a two-hour operation. I stayed with Heather all day, and Janice sat with her all night. During this three-day visit, we looked at possible retirement sites for ourselves before returning home.

We now had lived in the big house and yard on the bank of Bayou Lafourche for thirty years. It was increasingly expensive to maintain, contained a lot of unused space, and needed more personal attention than two aging seniors could continue to give it. Also, my back problem was forcing me to realize that I did not want and could not continue much longer the struggles of the last two years. Janice didn't want to stay in Thibodaux in a significantly smaller home. We briefly considered Monroe or West Monroe, Louisiana, close to Jan and her family, but did not feel encouraged in that direction. Janice's mother, sister, and aunt were in Raleigh, North Carolina; but that's a two-day drive from Thibodaux where Dan and his family would probably remain. Nashville, Tennessee, was the home of our second daughter and was almost equidistant between Raleigh and Thibodaux. A new Mormon temple would soon be erected somewhere in the Nashville area. Josh was an outstanding basketball player who would be a high school varsity starter in ninth grade there. We wanted to watch him play. And so our game plan became one to salvage as much as possible from our property and business before moving and to do so without unfairly handicapping our son Dan.

For the rest of 1997, I felt like a chicken running to and fro dodging traffic on a busy road. On the other hand, Janice was happiest when the house was clean and her wheels were rolling. October 10, she drove to Raleigh to be with her mother and Aunt Chloe, who underwent cataract surgery. Neither of the ladies drove any longer and depended on others for transportation. She called me to fly there at the end of the month and drive home with her later. We toured new subdivisions for prospective homes, in which her interests always focused on houses and yards larger than I thought we could maintain. As a past-due birthday present, she treated me to a massage by an attractive masseuse who might have talked me into almost anything but another big house and yard. We left for home November 2 driving to Helen, Georgia, a scenic Bavarian tourist town, where we spent the evening before arriving in Thibodaux the next day.

Later in the month, Janice drove to Nashville arriving early for the Thanksgiving holiday. Three days later, on a Monday, she met me there at the airport, and we went to look at houses and condominiums. Dan and Christina arrived from Thibodaux about midnight Wednesday. The ladies cooked Thanksgiving dinner while the young men golfed and the old man watched TV after a short walk. The girls spent the next two days shopping and looking at houses while the golfers golfed. The old man drove a golf cart on the last day and enjoyed the evening with family, visiting and bonding with grandson Jake and Rudy, the cat. November 30, we departed for Thibodaux and home.

In early December, we drove to north Louisiana to be with Jan and family as they welcomed Holly home from her eighteen-month church mission. We also spent two days and evenings in New Orleans at the Wyndham Riverfront Hotel as guests of the Zeneca Corporation. Sugar cane consultants were there for a seminar. The ladies were there for fun. On the first evening, Janice dined with a group at Antoine's Restaurant while I stayed at the hotel to watch the Tennessee-Auburn football game. The next evening was more romantic. And the next morning, we dined on beignets and café au lait at the Café DuMonde before walking in the flea market and shopping at the Lakeside Mall.

We celebrated an early Christmas dinner Thursday, December 18, at Bubba's Restaurant with Dan and Christina and her parents, after which presents were exchanged. Saturday, we drove to West Monroe with presents for Jan's family where I began to catch cold. Sunday, we arrived in Nashville bearing gifts and for a brief visit with Nancy and Danny. I went to bed early with a cold. Monday, we left early for Raleigh and a twelve-hour drive, arriving there very tired and sick. Tuesday and Wednesday, I stayed in bed with a cough and diarrhea. On Christmas Day, Carolyn and Donnie, Ricky and Miriam with daughters, Kirstyn and Hailey visited from 11:00 AM to 4:00 PM for dinner and for the ritual opening of presents. I ate for the first time in two days some soup, toast, and a cookie. So much for hygiene and nutrition. For the rest of the year, I alternated between the bed, the rocking chair, and the toilet. By Wednesday, December 31, everyone in the house was sick with colds and whatever else I had spread around.

Final Years of Fieldwork. After watching the New Year's Day bowl games and taking down Christmas decorations, we left Mom Rogers and Aunt Chloe both sick, but hopefully on the mend, to arrive in Thibodaux and at home Sunday afternoon. Holly came to visit her grandparents, and the day after her arrival, January 8, we hopped in the van and headed back to North Carolina to take care of Mom Rogers, now in ICU at the hospital, and Aunt Chloe at home with pneumonia. Cell phone calls from Carolyn and Ricky suggested that we might not get there in time to see Mom before she expired unless we hurried. With this thought in mind as we made our way there, our Windstar van overheated and coasted to a stop on the interstate highway north of Atlanta. Four hours later, with a new radiator and $881 less in the bank, we continued on toward Raleigh.

Three days later, I left the girls to care for the sick and flew home to attend crop-planning conferences with customers and parish meetings with farmers. More than a month later and after nursing the sick back to health, Janice and Holly returned to Thibodaux. I paid a $100 fine for speeding fifty-six mph in a forty-five-mph zone to get to a client conference and spoke on a crop production panel at the annual meeting of the American Society of Sugar Cane Technologists in February. Then in mid-March, Janice flew back to Raleigh to be with her mother during surgery to implant a heart pacemaker. Meanwhile, I cut grass, attended conferences with customers and weekly reading lessons with Elaine.

I drove to Raleigh in early April to accompany Janice coming home. We visited the Biltmore Estate and spent the night in the

Biltmore Hotel before continuing to Nancy's home in Nashville the next day. We returned home after a three-day visit in Nashville, watching a video of Nancy and Danny's trip to Hawaii, and learning about their newly drawn will. They had signed a will before departing on their trip requiring that, in the event of their death, their children should be reared by a young couple of similar age in their church. In spite of their explanation that our age had been a large factor in their decision, we would have appreciated being included in the family discussions before the decision was made. Janice was offended and told them so, and we both believed that Danny's views of all faiths outside the Church of Christ were also a factor. However, life goes on and, since mine had already been in progress for seventy years, I knew that I probably would not be an ideal parent for a three-year-old boy and probably not for a twelve-year-old either, in spite of the fact that I loved them both very much.

Janice and I drove to Fairhope, Alabama, for a weekend in late May, a break for us both before the scouting season would start in earnest. We checked into the Grand Hotel on Mobile Bay on a Saturday afternoon and walked around the grounds before dining at the Bay View restaurant. I wondered how close we were to the site of the beach cabin where Mother and her preschool children had spent a week during a summer more than sixty years earlier. We spent the next day exploring and looking in the shops before dinner at the Grand restaurant. Monday morning, we checked out and drove to Foley for breakfast at Lambert's and shopping at the outlet mall where, at Janice's urging, I bought a blazer, four shirts, and three ties. Then we shopped at the farmer's market in Loxley before heading home.

As the busy season for field scouting approached, I realized how tired I had become of all the hassle and became increasingly aware of my physical problems. I saw the ENT specialist at Ochsner Clinic about breathing problems caused by my deviated septum, which Heather likely had inherited from her grandfather. My ophthalmologist prescribed a salve and hot compresses several times daily for three weeks before surgery on an eyelid, and later the glaucoma pressure in my right eye got temporarily out of control. My physical therapist estimated 60 to 80 percent loss of strength in several leg and foot muscles from my earlier back surgery and suggested regular exercises for my left ankle and leg. As problems began to multiply, it became difficult to get access to my principal care physician (PCP). So I called the Ochsner-65 HMO office and told them to "get lost" and returned to Medicare. Next came an amazing array of expensive tests at Thibodaux General Hospital: an MRI, myelogram, CAT scan, and EMG just to diagnose hand and wrist pain as carpel-tunnel syndrome. These were followed by carpel-tunnel surgeries, first on one hand and then the other, during the busy 1998 July and August season. After healing from surgery, the pain was still there. I later learned that much of the pain had been due to worn-out joints at the bases of my thumbs, which I have learned to live with so far by taking daily amounts of glucosamine, chondroitin, and methyl sulfonyl methane (MSM).

In the early spring, I had offered my grandson Denny a summer job, but he had not been interested. Then a farmer client, Pete Lanaux, had asked if I might have a summer job for his grandson Travis. Travis impressed me favorably in his interview as an honor student in premedicine and as one unafraid of dirt or sweat. I hired him on the spot. After a few training sessions with me and alternating trips to the fields with Dan, Randy, and me, it became apparent that Travis would be dependable and also liked and respected by the growers. Therefore, as quickly as possible, I began to let him work alone as I checked behind him frequently. Within a few weeks, Dan and Randy were heading out in one direction each morning and Travis and I in another. I was no longer able to move as fast and cover as much ground as I had done only a year earlier. However, recommendations were always signed by either Dan or me. Also, I missed going to the field entirely on some days due to hand surgeries (July 22 and August 12) or doctors' appointments. In spite of difficulties, we made it through another scouting season without a major problem.

Janice and I drove to Gulf Shores, Alabama, on Saturday, September 12, 1998, to spend the customary week at Orange Beach in the Summer House condominium, where we were joined by Nancy's family. One of the things I enjoyed most about beach trips was the opportunity to stay in the shade for an extended period. I visited the exercise room daily, read, watched TV, and fished a little with Josh. Danny and Nancy took us to dinner at Mango's Restaurant on the eve of my seventieth birthday where I was wined, dined, serenaded, and presented with new shirts by my wife and with Dr. Laura's book on the "Ten Commandments" by Nancy. After a week at the beach, we drove home in the rain.

To accommodate Janice's restless spirit, and in sympathy with Willie Nelson's "On the Road Again," we departed Thibodaux for North Carolina on Saturday, October 10, arriving in Raleigh the next day. Five days later, I flew home to tend to business, leaving Janice to visit her mother and aunt for three weeks. Much of my time was spent visiting clients and some of it in seeing doctors: my urologist for old men's benign prostate hypertrophy, my ophthalmologist for glaucoma and visual field testing, and my dermatologist for the regular annual removal of incipient skin cancers from sun exposure. Having completed these chores, it was time to fly back to Raleigh to accompany Janice to Pigeon Forge, Tennessee, for two nights at the Comfort Inn by the Apple Valley farm, an evening at the Governor's Palace theater, and trips to the winery, the creamery, and the Tanger Outlet Mall. Then on to Nashville to visit and to look again at the River Plantation condominiums in Nancy's neighborhood. Finally, on Friday, November 6, we arrived home in Thibodaux after dark.

Nancy was unable to get home for Christmas Day, but she and Dan and their two families shared a late Christmas dinner with us at home in the big house on the Bayou for the last time Friday, December 27, 1998. Nancy stayed to visit for four days during which Josh helped Dan at his new house, and we dined together at Bubba's on the Bayou on the eve of New Year's Eve. Dan was so busy with home building and helping with their new baby that Randy did most of the soil sampling work during December.

We began the New Year of 1999 taking down Christmas decorations and eating the traditional New Year's dinner of cabbage and black-eyed peas. I had determined that this would have to be my last year of active full-time consulting but had not decided upon the future of the corporation. It would have been my dream come true to turn day-to-day operations over to Dan, with a college degree in agriculture and a trusting clientele. Dan was an experienced and competent scout for dealing with the major insect problem on sugar cane, and this alone would satisfy many of our customers. However, he did not have the prerequisites in education, training, and experience, now widely considered essential, to deal with the variety of problems and events that may arise from time to time and to command the respect of a great majority of clients. Some growers were concerned about a circulating rumor that he had an addiction problem, and I was worn-out from constantly arguing with him about endless details of my business operation. Dan was being paid a monthly salary, but he was so busy working on his new house that Randy did most of the soil sampling this winter.

I wrote a check in January for down payment on a condominium under construction in a senior retirement community in Franklin, Tennessee, sixteen miles south of Nashville. Later that month, we attended Steve Hoak's wedding and reception in Quincy, Florida. Three days later, I attended an afternoon meeting of Lafourche Parish sugar cane growers to hear research and extension people talk about their studies and recommendations for the coming season. At the end of the month, Janice drove to Nashville to attend a baby shower for Nancy and to pick out paint, paper, floors, and light fixtures for our new condo.

Meems had been suffering a variety of age-related problems for months when Judy offered one morning to take him to the Ridgefield Hospital for animals. Not wanting my secretary to suffer that heartbreaking chore, I took Meems to the hospital myself and went back for him after a couple of hours. His body was still warm, and I felt guilty for not staying with him to the end. I buried him at the foot of a hackberry tree on our bayou bank approximately seventeen years after twelve-year-old Dan had brought the kitten home with wide-eyed enthusiasm. A neighbor had given it to him as a present, and we didn't have the heart to refuse it although we had raised, loved, and lost many pets before. We had mistakenly identified the sex of this small kitten as female and named it Mimi. As our error became increasingly obvious, the name was increasingly inappropriate. But how can a name be changed without trauma when everyone, including the cat, has become accustomed to hearing it? Meems was the obvious answer, a name that stuck and still recalls fond memories. Two days later, we signed on the dotted line to accept the tendered offer from the buyer of our dream house on the bayou.

On a Friday evening in mid-February, Dan and Christina visited to express their surprise and dismay that I should be contemplating the sale of my corporate consulting business. They had heard that I had discussed this possibility with the owner of another consulting firm. I believe that Dan felt he was being double-crossed by his father. I saw it in a very different light. If another consultant could fairly reimburse me for some of the thirty-five years of effort spent building this business, hire Dan at better than his present salary, and provide an environment in which he would be encouraged to continue his education and improve his knowledge and training, then we might all be long-term winners. As it turned out, and not surprisingly under the circumstances, no one was interested in buying the business.

In late February, Dan, Randy, and I attended the annual meeting and workshop of the Louisiana Agricultural Consultants Association at which Randy was presented and accepted for associate membership. Randy had developed a real interest in agricultural consulting and seemed anxious to pursue it. Dan had been voted into the association the previous year in absentia. Later that month, I signed a one-year lease on a rent house to double as my place of business and as an "off-and-on" residence with a front office, kitchen, lounge, bedroom, and bath. This upset Dan who thought that the office should have been moved into their three-bedroom home with him and his family.

In early March, Janice drove to Nashville to help Nancy following the birth of our newest grandchild, Jillian Grace. I stayed home and began moving the office from the big house to the rent house across the bayou at 1648 Highway 1. We moved gradually over a period of several weeks during which our children helped pack boxes, conduct a garage sale, and distribute furniture and household items, which we would no longer use, among their various homes. Judy's secretarial and Carrie's housekeeping skills now included packing silver, glassware, china, and knickknacks. With a U-Haul trailer, we delivered most of our family room furniture to Jan in West Monroe although our buyer had offered to purchase it for several thousand dollars or for whatever we thought it was worth. As there would not be room in our new condo home, I also gave away much of my private library, including forty-six volumes of the *Annual Review of Entomology*, valued at more than four thousand dollars. We were not wealthy now and never had been. However, Janice was suffering such emotional trauma from having to give up her dream house that I encouraged her to furnish the new condo as she wished. Consequently, upgrade floors, carpets, and drapes as well as new furniture were purchased without much restraint. I thought that we might live to regret it but couldn't imagine her living happily otherwise.

An Atlas Van Line truck loaded the rest of our furniture and belongings at the big house on April 29 and unloaded them into a Mallory Station Storage room in Franklin, Tennessee, May 1, 1999. In spite of the confusion associated with moving, consulting activities proceeded on schedule, including the scouting of winter wheat and early spring-planted corn and the preparation of field maps for sugar cane and soybean scouting. Some late soil sampling was also accomplished. In late May, I began to have trouble again with my back and my right leg this time. On May 26, I was barely able to get out of bed for bath and toilet. I walked with crutches again or hung on to every available thing for support for a while. Some days were better than others when I would walk with a cane or even without any extra support.

Friday, May 28, Dan, Randy, Travis, and Judy all worked on field maps. The radio tower was reactivated for our office, and truck radios and cell phones were given to each fieldworker. A drug policy was read and signed by each employee by which they agreed to drug testing at any time and to dismissal from their jobs on the basis of positive test results. I was no longer able to be fully responsible for field scouting. All sugar cane farms were assigned to one of five routes, theoretically based upon five working days. The farms on each route were divided among the three scouts so that, in the event of an error, there could be no doubt about who was responsible. Dan disagreed with this policy and felt that it deprived him of the ability to keep in touch with all the clients. Thus, cane scouting began in earnest in late May, a little earlier than necessary, to give everyone time to get well-acquainted with their farms and fields.

I flew to Nashville for a Sunday walk-through of our new condo before writing a final check at the closing ceremony June 17, 1999.

Immediately afterward, Janice drove me to the airport for my return flight to New Orleans and three doctor appointments at Ochsner Clinic the next day. She returned to Thibodaux in late July to spend some time with me and to purchase an imitation fiberglass sugar pot, similar to the real cast-iron one we had left as a planter in the front yard at the big house. We placed the new sugar pot in front of the rent house, filled it with water, and installed a pump and watertight lamp in it to create a decorative fountain that we would later take to Tennessee. She then bought several flats of flowering plants that we planted around the house and fountain.

I did a little bit of scouting when I felt that I could, but generally followed behind to visit with growers or deliver recommendations, attended the league's contact committee meeting in mid-July, and drove with Janice a few days later to Panama City, Florida, to watch Josh play baseball for three days with the Nashville Sabres in a tournament of thirteen- and fourteen-year-olds. Janice drove to West Monroe toward the end of the month to visit Jan for a couple of days, while I attempted some catch-up scouting, and Dan played softball in New Orleans. My back problems emerged again August 9, prompting me to go back to physical therapy for the rest of the month, but without much effect. Dan, Randy, and Travis continued their regular scouting schedules through mid-August, after which college classes started for Randy and Travis. Randy came to help out during free time and on weekends until we finished the season.

Janice and I celebrated our forty-sixth wedding anniversary in Baton Rouge September 11 by buying a new chest with a black marble top, two lamps, and a mirror for the entrance foyer of our new home in Tennessee. I gave Dan a letter on September 15, 1999, terminating his employment but with a separation package amounting to the equivalent of his salary for the remainder of the year. At the same time, a letter was sent to customers with the following message:

> Back problems, which have continued since my surgery in 1996, have forced me to be less and less active in the field. I have had to refrain since mid-August this year from most field activities and at this time will apparently be unable to resume them. Therefore, I am retiring from active field work after 43 years during which I have worked with sugar cane and sugar cane growers in Louisiana. I very much appreciate the opportunities you have given me to serve these many years as well as your friendship, trust, and respect. Thank you much!
>
> My son Daniel will continue in this ag service business. He has worked with me for 8 years as a field scout (1992-99) and as a certified consultant in entomology from 1996 to present. He has the ability and experience to monitor insect populations in sugar cane and to recommend control measures when needed. Also, he has been responsible for soil sampling during this time. I hope that you will consider using his services. He will be contacting you in the near future about this.

We drove to Franklin the following day to meet the movers at our storage facility and to direct moving of the large items of furniture into our condo at 8002 Sunrise Circle in the Morningside senior retirement community. Numerous boxes of clothing and other items remained to be slowly removed from storage during the next three years. Near the end of the month, I returned to Thibodaux and employed Randy to help me finish scouting a few fields of soybeans for another couple of weeks. Actually, I sat in the truck, and Randy did the legwork to complete my last season of crop scouting.

In late February 2000, I gave Dan my office desk and chair and a couple of bookcases. The rest of the furniture was loaded on to a U-Haul truck, which I drove as Janice followed in our van from Thibodaux to Franklin, Tennessee. We were making good time through Mississippi in cold and rainy weather when a loud noise startled me. I pulled off on to the shoulder of the road and stopped. There were pieces of glass all over me and the empty seat beside me, as well as the floor and dashboard, and a strong draft of cold air had blown on the back of my head. The rear cab window of the truck had exploded into thousands of pieces covering the interior of the cab. It was Friday and late afternoon when most businesses were closing. I got out and checked the tires, and we decided to keep on trucking. Janice gave me a blanket from the van to wrap around my head and shoulders, and we drove on. From this point on, every mile was agonizing as wind and rain whipped my shoulders from behind, and I struggled to keep the blanket in position to protect me as much as possible while driving. On I-65 north of Cullman, Alabama, traffic was heavy, but moving fast in bad weather, when I hit and crushed a large dog crossing the road. Janice saw it in her headlights and called on our walkie-talkies to say that she was so shaken and tired that we needed to stop and spend the night. Without her wise suggestion, I would have been foolish enough to drive another hour and a half to Franklin. Fortunately, a motel was in sight at Athens, Alabama, where we stopped for the night. Two days later, Josh and Danny helped unload the truck at our storage facility, and I swore, for what I hoped was the last time, that I would never ever move again.

17

Retirement and Retrospection

Better Days. Christina Adams may be the best thing that ever happened to our son Dan. During the years following their marriage, she helped him to beat a drug problem that his parents were ignorant of in the beginning and later helpless to solve. They now have two happy and healthy children, Dallin Henry and Delaney Rose, whom they both adore and who adore them. Dan loves his family dearly and, in spite of his father's often poor example, devotes quality time and attention to them in ways that are heartwarming to see. He retains a significant portion of the consulting business from which I retired, works at other jobs as well, and shares with household chores while Christina helps with the home office and is a legal secretary for a local law firm. Dan continues to golf and play baseball himself and coaches Dallin in his team sports. He has a big heart, would give the shirt off his back to anyone in trouble, and spent most of his time for days and weeks assisting and befriending evacuees following the Hurricane Katrina disaster. We are unable nowadays to see them as often as we would like.

Seen even less often is our daughter Jan, who at last report was working happily as a teaching assistant (paraprofessional) in the West Monroe High School, and her husband Bob, now graduated in education from the University of Louisiana at Monroe and teaching high school history in a private school there. Their three oldest children are all married and living in Texas with two great-grandchildren; the next two daughters are attending college in Monroe and the youngest son is in fifth grade.

What If. If upon retirement I had been assigned a private cubicle in the Ellender Memorial Library, I might have been tempted to write the book on "Sugar Cane Insects of the World" that the LSU Foundation had commissioned me to do a year before I left LSU and thirty years before my retirement from Nicholls. I had made a very small beginning at such an effort by the private publication of a bulletin, entitled "Basic Information on Sugar Cane Pests" in 1991. However, time and circumstances did not permit my continued indulgence in such activity.

There is little profit to be gained in the late afternoon of life by speculating about "what ifs." However, these thoughts do occur briefly from time to time. What if I had not jumped those ditches on an ATV? Would I have needed back surgery? What if I had not branched out into household, structural, and horticultural pest control services? What if I had resisted the temptation to get involved with RSD testing? What if I had not resigned from the LDS church branch presidency to go to Egypt? What if I had accepted a job offered at Virginia Tech instead of LSU, or had remained at LSU without going into private consulting, or had accepted a job offer from the University of Illinois, or had investigated an invitation to apply at North Carolina State instead of remaining at Nicholls? What if I had been more diplomatic, more socially inclined, and less confrontational? What if I had spent more time with family and less time at work? Some things could have been better. Many things might have been different. But "we know that all things work together for good to them that love God . . ." (Romans 8:28).

Hall of Fame. While packing and preparing to give up my office in Thibodaux in late February 2000, the phone rang, and I was surprised to hear Calvin Viator say, "Henry, I have something for you at my house if you would like to come by." I said that I would and would also bring him a box of reprints of our published ant research that he might be more likely than me to have use for in the future. Janice and I stopped by for a short visit and farewell to Calvin and Jeanie, upon which he handed me a handsome plaque that now hangs on my study wall. The upper left portion of the plaque is a relief map of Louisiana to the right of which appear in large letters "LOUISIANA AGRICULTURE HALL OF FAME." At the bottom edge and beneath a small scroll are the words "LOUISIANA AGRICULTURE CONSULTANTS ASSOC." Above the scroll, the plaque reads as follows:

Presented To

Dr. Henry Long

*For Your Years Of Dedicated
Service To The Betterment Of
Louisiana Agriculture
February 15, 2000*

I was genuinely surprised and had no idea that anything of this sort was being considered. I had not attended the annual February meeting and workshop for the first time in many years because I was packing to move. I felt very grateful and honored and was sorry I had not had the opportunity to express my appreciation in person at the meeting. I was consoled by the thought that some others had been so honored posthumously, so I was not first in being unable to personally express thanks. Calvin had been secretary and treasurer of the association for many years until recently when Cecil Parker had assumed that responsibility. He said he had told Cecil to be sure that I got to the meeting, but I never got the message. I wrote Cecil a letter of thanks to the association, asking him to read it to the membership at their next meeting. I have since been told that the names of these hall-of-fame members are displayed on the wall outside the office of the Commissioner of Agriculture in Baton Rouge. I haven't seen that but do appreciate the honor.

Life in Tennessee. My retirement from agricultural consulting began officially January 1, 2000. Our condominium in Villages of Morningside, A Senior Living Community, Franklin, Tennessee, is sixteen miles from Nancy and her family in Bellevue, a west Nashville suburb. We had considered other possible locations in or around Bellevue but settled on Morningside instead because of its proximity to excellent shopping, the nearby LDS chapel and temple, and the apparent uncertainty of our daughter's husband about where they might eventually move. We both now certainly were qualified by age, to live in this senior community. Giving up the big house in Louisiana, which we no longer needed had been a tough sacrifice for Janice to make. For that reason, she had a free hand in furnishing and deciding on upgrades for our new home. Determined to put as much as she could into this little condo less than half the size of her dream house, she spent another 20 percent on upgrades and landscaping. When we had enquired at the sales office about association rules and regulations governing landscaping in the common ground, which includes all land outside building walls, we were told not to worry about it so long as what we did was attractive and reasonable. To insure that what we did was in good taste, we employed a nursery and landscape architect at considerable expense to design and plant beds with shrubbery and flowers.

As soon as we had moved in and started to get settled, we realized that our lighted fountain and our enlarged and added planting beds were cause of no little attention, which included the admiration of some and concern of others. This prompted me to begin to study the association's bylaws and rules and regulations carefully for the first time. We had received copies of these hefty documents only at the closing of the sale when the builder's attorney, Ms. Pigg, paused a few moments to ask if we had any questions.

Following the first meeting of our association of owners in September 2000, the newly-elected president of our board of managers, Jim Ratcliffe, asked that all the ladies meet briefly with the board secretary, David LaFond, who would discuss with them some of the rules and regulations about landscaping and decorating outside the buildings. The ladies were told in a rather forceful manner that any exterior decorations of the building and additional plantings, beyond the few green shrubs provided by the developer in the fronts of the units, could be done only with written permission from the board and following a written request to the board. I was told that Mr. LaFond further clarified his meaning by saying, "and ladies, if the board gives you permission to plant twelve daylilies, that doesn't mean thirteen." A number of ladies, including my wife, left the meeting with distinctly ruffled feathers.

Jim Ratcliffe was a retired accountant from the Vanderbilt University business office who often walked his dog in the afternoon as I watered my enlarged and added landscape beds and plants, checked my fountain, and sometimes tried to engage him in conversation. He showed many symptoms of the typical CPA syndrome with the usual compulsion for crossing t(s) and dotting i(s). Unlike many CPAs, he had the aura of one with a distinct superiority complex. I would hear him comment on a later occasion that "some of the dumbest people in the world were professors with PhD degrees," which of course is sometimes true but not the best way to win friends and influence many people that I have known. He wasn't very communicative, and when I prodded for his opinion about how far was too far in landscaping, he would simply say that everything was up to the board. However, it soon became apparent that Jim pretty much controlled the board's thinking as well as its actions. I began to wonder if, under his influence, they might require us to remove our already-expensive landscaping at considerable additional costs and to the added unhappiness of my wife. Therefore, when the board called for volunteers to serve on an architectural and landscaping committee in January 2001, I volunteered to serve along with half a dozen other residents and two board members.

At our first committee meeting, there was discussion of the fact that a number of residents had placed bird feeders, birdbaths, statuary, flags, and other objects in front of and between units, which would need to be moved before the mowing season started. There was also concern expressed by some about such items that had been placed among the shrubs in the original front beds, but without permission from the board. It was also noted that some residents had enlarged their front shrubbery beds to plant flowers and some had even created new beds along the sides of their units without board permission. Never mind the fact that these things were attractive additions to the community, we seniors and one old maid, Peggy Lawrence, had found something new to fret about. In fact, Peggy was so outraged by the size of my added flower bed with its lighted fountain that she believed I should be made to return the common ground to its original state. I pointed out that the difference between my beds and some others, which had been added or modified without permission, was only a matter of degree, and I challenged her to find anyone else who would say that our landscaping was not attractive. The meeting ended with committee chairman and board treasurer, Dave Boston, trying to calm Peggy down in a corner of the room as the rest of us left the building.

Subsequently, another board member, Hunter Neubert, was appointed chair of the landscaping subcommittee. Hunter was a diligent, well-meaning, and hardworking board member whose wife met regularly with the board as the voluntarily unpaid editor of the *Morningside Monitor*. After meeting another time or two to discuss landscaping guidelines, Chairman Neubert wrote a one-page set of guidelines that became an official part of our association's rules and regulations. In essence, these rules did not require removal of landscaping already done but urged residents to try to maintain "the type of landscaping that has been established." A later clarification of rules further stated that "beds planted by residents are the resident's responsibility." Later, the landscaping guidelines were again tightened and restated in eight numbered paragraphs that included rules against any ornamentation in front beds, any front porch furnishings or decorations not approved by the board, and any new flower beds not approved by the board.

In 2001, I sent a newsletter to every resident of the association urging them to be on guard against the dictatorial rule of a board bent on exercising its authority instead of promoting harmony and the general welfare of our community. I then circulated a petition, which was signed by about three dozen out of one hundred thirty-six residents, requesting that the board be guided primarily by consideration of the principles of "protection of property values" and "avoidance of unwanted maintenance costs" in dealing with matters of rule compliance. A number of others said that they agreed with the petition in principle but didn't want to get involved by signing anything. My point was that many rules were not enforced and need not be enforced unless the board deemed that they threatened property values or maintenance costs. In that way, the efforts of some to beautify the community by planting, decorating, and furnishing common areas could be celebrated rather than penalized when these efforts did not threaten community values.

I volunteered and worked for a year on the bylaws committee that revised and rewrote that document. In September 2002, I was nominated to replace a board member whose term had expired but failed to get enough votes for reelection. I later learned that an influential board member, who had urged me to run, had later done "everything possible" to prevent my election. In April 2003, I joined the board at its request to serve five months of the unexpired term of Bob Gaunt who was forced to resign for health reasons. In May, Jim Ratcliffe rejoined the board temporarily, at its request, to replace a member who had moved away. In planning for the approaching September meeting of the association, Jim wanted to institute an oath of office to be taken by newly elected board members at the meeting. By this oath, each newly elected member would pledge to uphold the "bylaws" and "rules and regulations" of the association. When I questioned the need to do this with new members only when old members had never taken such an oath, Jim refused to discuss it further, and I refused to take the oath. Oaths were not discussed thereafter.

Prior to Jim's temporary return to the board, I had placed a round white aluminum bench, which had decorated our front yard in Louisiana for nearly twenty years, around an oak tree in the common ground between us and our neighbor, Camille Snow. It was beautiful and useful, and Camille and I both rested on it during occasional breaks from working in our flower beds. The feet of the bench rested in the mulch around the tree and posed no problem for lawn maintenance. I had not gotten written permission from the board but had showed it to the board members present at an earlier meeting and asked if any were opposed to my placing it in the common ground between me and my neighbor. They had admired it, and none had expressed opposition to my proposal. However, an influential board member, Christine Bess, later received two phone calls from two residents who wanted to know "what other rules the board would permit Henry Long to violate now that he is a board member." Christine raised the issue at our next board meeting, and after some verbal arm-twisting by Jim Ratcliffe, the board voted, with me abstaining, that I must remove the bench from the common ground. I complied that very day. Reasons given for this enforcement were: my failure to obtain written permission and telephoned complaints from two of 136 community residents. Our board's authority is over Morningside-1. The boundary between Morningside-1 and Morningside-2 is about ten steps from my front door. Several weeks later, I obtained permission to place the bench around a large tree thirty paces from my front door and in the common ground of Morningside-2. It presently serves to beautify the entire area and provides occasional rest for workers, walkers, and even practicing golfers. Looking at it from my front door also provides a little salve for my wounds.

I allowed my name to be placed in nomination once again for the September 2003 election of board members when three of five seats were vacant. I came in fourth in a field of seven candidates. I did succeed in influencing the board for the first time to allow the numbers of votes cast for each candidate to be made public and in changing the form of the proxy ballot to permit the residents who give their proxies to vote their own choices instead of leaving this decision to the discretion of the holders of the proxies. My statement published in a summer edition of the *Morningside Monitor* read as follows:

> I am a nominee for the Morningside Board of Managers because I would like to promote more communication between the board and co-owners, greater participation of co-owners in association matters, and a reasonable philosophy on rule compliance.
>
> Governing documents exist primarily to protect property values and to control maintenance costs. I

support the reasonable interpretation of by-laws, rules, guidelines and all governing documents. I believe that their enforcement should always be based upon the protection of property values, the control of maintenance costs, and concern for the well-being of co-owners. Prudence suggests that we obtain Board permission to do certain things; however, failure to seek permission should not alone be cause for rule enforcement.

Appearances affect property values. Plantings or furnishings in common areas should complement or enhance and not detract from original patterns of architecture and landscaping.

The best government of community associations occurs when the maximum numbers of co-owners communicate their opinions on association matters and are willing to participate in the governing process. The alternatives lead to government by a few who decide what's best for all.

Following the 2003 association meeting, I declined to serve on a newly formed insurance committee and decided that my obligations to the association had been met, at least for a while.

I presently attend Mormon worship services with Janice at the Carnton ward of the Church of Jesus Christ of Latter-Day Saints. Although many years have passed since I gave up my membership in the LDS Church, I feel a great affinity for its people and for many tenets of their faith.

Since moving to Tennessee, we attend sports competitions and other events in the lives of our nearest grandchildren and sometimes baby-sit for their parents to globe trot. We have enjoyed and benefited much from exercising at a nearby gym, which we do most days. Other time is spent reading, gardening, at the computer, attending the Tuesday morning men's coffee, paying bills, resisting sales gimmicks directed at senior citizens, contributing to charities, breathing, sleeping, eating, supplementing, medicating, urinating, and defecating. And all of these require much more time to do now than they did just a few years ago.

Revisiting Nicholls State. While visiting my son Dan and his family in Louisiana during Thanksgiving 2004 and in the summer of 2005, I was privileged to visit several faculty members and friends from earlier years and to learn that Dr. Ayo retired from the university in May 2003 and was replaced by an out-of-state person, Dr. Steve Hulbert, with significant university administrative experience. From several reports, I understand that Dr. Hulbert expects deans and other administrators to do their jobs and to make their own decisions without much interference from his office. I am also told that there has been an accompanying major improvement in faculty morale. Perhaps now NSU may gain the respect that it deserves as an important regional university in the Louisiana system and as one that encourages excellence and discourages mediocrity.

Time for Genealogy. Now that I have almost finished writing these memoirs, I feel a desire to rearrange and add to, if possible, the information in Mother's book on "The Longs, The Listons, and Some Related Families." This will involve construction of pedigree charts, family group sheets, and individual records with the help of computer software developed by the LDS Church, additional genealogical records from the church genealogical library, and other historical information that may be accessed on the Internet.

My first addition to this knowledge was found on the Internet in a slightly edited version of an article published in the *New England Historical and Genealogical Register* (volume CXLIX, October 1995, pages 374-378) by Mathew J. Grow. As has my mother, so also Mr. Grow indicates that William Pratt, later known as Lt. William Pratt (the Settler), and his brother John came from England to Hartford, Connecticut, in 1636. An interesting detail, gleaned from this article and one that Mother's book does not mention, but that may be of interest to some of my children and theirs, is that this Lt. William Pratt is the great-great-great-great-grandfather of Parley and Orson Pratt, early leaders and apostles of the Church of Jesus Christ of Latter-Day Saints (Mormon) and that this same Lt. William Pratt is my mother's great-great-great-great-great-great-grandfather through her own grandmother, whose maiden name was Sara Eliza Pratt. I don't believe that Mother would have been happy to learn that in this life, even though only less than one-half of 1 percent of her genes could have come from this common ancestor, but I do like to think that they may all be good friends now.

Class Reunions. Several months ago, I received a notice from a high school classmate, Evelyn Sharp, informing me that it is time again to mark my calendar for September 23 and 24, 2005. Our high school senior class has held a reunion every fifth year since graduation in 1945, but I have never attended. My high school and college yearbooks were disposed of years ago along with old magazines and newspapers. In most of the years past, I always have had work-related excuses for not being able to go. But now since retirement, reasonable excuses are more difficult to find. However, even the thought of attending this reunion makes me uneasy. Of course, I realize that such feelings are irrational, but they are feelings that I've had for as long as I've known about class reunions. Perhaps I should go now, even if mostly out of curiosity, and try to dispel some of these long-held and unhealthy "ghosts in my attic" or "bats in my belfry."

My peers may have wondered why my brothers and I permanently departed Decatur. They certainly should not be blamed for this. As my brother Needham explained it to one of his friends, "the fact that we grew up on a farm isolated from peers and pushed unmercifully for good grades . . . while living under a myriad of incredibly stringent religious rules for daily life, which Mother espoused and Daddy upheld with an explosive temper and an iron hand, all added to a suffocating home environment. Isolation from peers, further aggravated by economic limitations, seriously impeded the development of our social skills. As a result, we never had the strong attachments for peers and for the environment of Decatur that some others have enjoyed. We were simply not a significant part of that picture. It was no fault of our peers. It was just circumstances. But feelings are real, and thus, Decatur did not become 'Bre'r Rabbit's briar patch' for us. There were just too many unhappy memories."

Janice and I attended both the Friday picnic and Saturday evening banquet at which I spoke briefly about "Bre'r Rabbit's briar patch." Many went out of their ways to welcome us. We were so warmly received that I felt a little like the "prodigal son" of scriptural notoriety and renown. I am very glad that we went but sorry that my feelings, etched in memory from so long ago, are still so very real. I was a little nervous and uptight most of the time.

18

EULOGY TO AN OLD FRIEND

My phone rang Friday morning, January 28, 2005. Upon answering it, I heard the voice of Norma Hensley telling me that her husband and my longtime friend and associate, Sess, had passed away the night before in his sleep at the age of eighty-five. He had battled diabetes for a number of years and had recently begun to suffer from Alzheimer's. Since my move to Tennessee five years earlier, we had continued to remain in touch by e-mail and an occasional phone call. The passing of one who had been such a friend and fellow worker for almost half a century came as a shock and with a sense of real loss and sadness. His funeral would be delayed for a week to allow two of his daughters to travel from their ice- and snow-bound homes in New York and Atlanta to reach Houma, Louisiana.

Within a short time, I decided that I would also travel to his funeral and sat down to write a eulogy to my friend in the event that I might be asked to say something. Janice and I drove to Louisiana on the following Friday and attended a beautiful, well-planned, combination military and church service the next morning. There was no request for me to contribute publicly to the occasion, and I probably would have been emotionally overcome in the process had I tried. In listening to the eulogies presented by children, grandchildren, and a couple of close family friends and neighbors, I was reminded of some important facts about his military service that I had forgotten. Upon returning home, I amended my efforts that I now insert here as follows:

> Sess Hensley was a decorated veteran and survivor of World War II with 82nd Airborne jumps over southern France and Italy before serving also in the Korean War. He was decorated with two Purple Hearts for wounds in action against an enemy and with a Bronze Star Medal awarded for heroic service in combat. After service in Korea, he retired from the military to earn a PhD in entomology at Oklahoma State University.
>
> I arrived in Louisiana in March 1957 as the newly appointed assistant entomologist in charge of sugar cane insect research at LSU. I first met Dr. Hensley in December of that year when he was a newly appointed entomologist at the USDA sugar cane research station at Houma, Louisiana. This meeting was the beginning of years of cooperative effort and friendship. We didn't always agree, but we always respected each other's opinions and contributions toward our common goal of improving the management of sugar cane insect pests.
>
> Radically changing insect control practices in Louisiana sugar cane during the late fifties and early sixties required a great deal of research as well as extension and public relations work. Sess had much more skill in the latter area than I. He was able to relate to people one-on-one in a positive manner, which I have long tried to emulate but never quite achieved. He enjoyed mixing and mingling with

folks and did it effectively, which, in addition to his research, was highly beneficial in the course of our work with farmers, plantation managers, and mill operators. I was a workaholic and not much at mixing and mingling. However, we regularly compared notes and benefited mutually from our respective observations as did also our efforts to promote more enlightened pest management practices.

By 1963, Sess had tired of battling his USDA boss over research principles and had moved to LSU in Baton Rouge to head the soybean insect project. We were neighbors there on Wylie Drive for a couple of years until I left LSU in 1965 to start a sugar cane crop consulting service and a nine-month teaching appointment at Nicholls State. Sess was then transferred from soybeans back to sugar cane where he remained for a dozen years conducting research and supervising graduate studies before retiring as a professor emeritus and returning to the USDA sugar cane station in Houma in 1977 as research leader for crop protection.

Over the years, we sometimes attended meetings together and discussed problems of mutual interest. A time or two, we even confronted each other in public debate over an issue about which we disagreed. However, our mutual friendship and respect always remained firm. In September, 1993, we traveled together to Miami, Florida, to be honored at the Inter-American Sugar Cane Seminar with plaques of recognition in appreciation of our "outstanding work in pioneering, developing, and practicing the environmentally sound practice of Integrated Pest Management." In 2004, Sess was further honored by induction into the Louisiana Agriculture Hall of Fame.

His friendship and contributions will be long remembered by all who knew him, and his work will be cited for years to come. His wife, Norma, and daughters, Sessalie, Lisa, and Sara, are justifiably proud of the recognition he has earned and the many contributions he made. He will be missed by all who knew him.

19

Religious Faith

My earliest childhood memories include regular weekly church attendance and summer Bible school at the First Presbyterian Church. Our Bible school attendance was limited to those years before we moved to the farm. In addition to the Long boys, the children there included Dan and Neal Speake. Dan is now an emeritus professor and retired director of Wildlife Research at Auburn University. Neal was my age but a year behind me in school and is now deceased for almost fifty years. Also present were A. J. Coleman, now a practicing attorney in Decatur, if not retired, and Barrett Shelton, now editor of the *Decatur Daily* newspaper as was his father before him. Little girls included Jean Link, mentioned earlier in the swimming pool episode, and Mary Emily Burnham who lived across Ferry Street from the Speake boys. There were others whose names and faces I might recall with prompting. Teachers included Mary Louise Garrett, an attractive young lady schoolteacher, whose father owned and operated the local dairy and ice cream plant; Mrs. Shumake, whose husband, Claude, owned a local furniture store and also taught the young boys' Sunday school class; our mother; and some other mothers as well.

We attended Sunday school and worship services every Sunday of every year, unless serious illness prevented, as long as I lived at home with parents and brothers, which was until four months before my seventeenth birthday. It would be difficult to say which seats we boys squirmed in the most. Neither the small wood chairs in the classrooms nor the long wood benches in the sanctuary were padded, and there was little natural padding on my bottom. Our family pew was third row from the front and left of the left-hand aisle, and there was always the same lady with the same hat sitting in front of me potentially blocking my view of Rev. Wallace if I had wanted to see him. Family names were not engraved on pews, but ushers knew where to seat visitors and where not to seat them. It was not unheard for an elderly, longtime lady member to occasionally inform an unescorted visitor, who might wander into the wrong pew, that he or she was in the wrong place. When very young, we always sat with our parents in the family pew. As we became older, we were sometimes permitted to occupy a pew with the Speake boys at the back of the sanctuary. Of course, whispered conversations, giggling, and other activities were more or less continuous in that pew, unless or until the preacher interrupted his sermon to call us down, which he sometimes did. When this rarely happened, the privilege of sitting with Dan and Neal was lost for a while.

It was not easy to regularly attend the Wednesday evening prayer meeting after moving to the country due to the difficulty and distance of travel and numerous farm chores. However, we made up for this failure by organized family singing of hymns and memorization of scriptures on Sunday afternoons, after which, if there was still daylight, we could play outdoors. Also, we were required to study the Shorter Catechism with the goal of reciting it at one sitting to an officer of the church for which a certificate was issued. My certificate certifies that I memorized and recited at one sitting the Introduction to the Shorter Catechism April 16,

1939. The Introduction included the first thirty-eight of one hundred seven questions. These thirty-eight questions deal with what man is to believe and the remaining sixty-nine with how he is to live. At my present age of nearly seventy-six, most of the one hundred seven sound familiar, and I can still quote verbatim answers to a few of the first thirty-eight.

Then as now the Shorter Catechism presented questions with answers that did not always satisfy, sometimes seemed dogmatic and are debatable based on scripture. For example: *Question 6.* "How many persons are there in the Godhead?" Answer: "There are three persons in the Godhead: the Father, the Son, and the Holy Ghost; and these three are one God, the same in substance, equal in power and glory." *Question 7.* "What are the decrees of God?" Answer: "The decrees of God are his eternal purpose, according to the council of his will, whereby, for his own glory, he hath foreordained whatsoever comes to pass." I now believe that these answers represent the collective judgment of honest and sincere men seeking to identify important points of doctrine and to interpret scripture to the best of their ability with as much inspiration as they could muster. I concede the same to the many authors, some unknown, of other parts of the Constitution of the Presbyterian Church (USA), viz the Apostle's Creed, the Scot's Confession, the Heidelberg Catechism, the Second Helvetic Confession, the Westminster Confession of Faith, the Larger Catechism, the Theological Declaration of Barmen, the Confession of 1967, and A Brief Statement of Belief of 1983. The differences among Christian groups in points of doctrine or emphasis thereon now seem trivial in comparison to the importance of the Lord's words: "Thou shalt love the Lord thy God with all thine heart and thy neighbor as thyself."

It should be remembered that the political goals and ideals of a single fourth-century man, the Roman Emperor Constantine, heavily impacted subsequent Christian belief and thought through the edicts of his arranged Council of Nicaea in AD 325, now known as the Nicene Creed. One may speculate about how much scripture may have been lost, altered, or deemed to be uninspired both during and resulting from the deliberations of this council as it sought to weave and trim the many strands of existing Christian thought and belief into a single universal Catholic Church. This church came to believe that it could control the presence or absence of God in people's lives through the administration of sacraments and forgiving sins by priesthood with sole authority to act on earth on God's behalf. Then, in 1450, the invention of the printing press opened the possibility to many more people reading the Bible directly. The ensuing Protestant revolution quickly changed many Christian concepts. No longer could the Catholics pretend to control the paths by which God comes to man. We began to realize that God was in charge, not the priests. They were only facilitators, without any real power to bestow or withhold God's gifts (Ed Duemler, *The TENNESSEAN*, page 10A, June 3, 2004).

I was most actively religious as a Latter-day Saint (LDS or Mormon) from my conversion to that faith in 1958 until shortly before my separation from it in 1975. Scattered accounts of that activity have been given here in earlier chapters dealing with periods from the late fifties to the middle seventies. Religion was not an important focus in my life from 1974 until 1986. However, in that year, I bought a new leather-bound second edition of the Revised Standard Version of the Holy Bible with my name imprinted in gold on the front cover. I began to read it daily and to pray again, becoming increasingly aware in the process of how true it sometimes is that if you "train up a child in the way he should go, when he is old he will not depart from it" (Proverbs 22: 6).

I began to attend regularly the First Presbyterian Church of Thibodaux where I was welcomed by a small congregation in a small building, which claimed to be the first Presbyterian Church established west of the Mississippi River. Protestants had long been a minority in this predominantly Catholic part of Louisiana. I attended a Bible study group and tried to be helpful or useful as opportunities presented themselves at church. Without aspiring to the job, my election to elder was announced in the church bulletin of October 31, 1993. Monthly meetings of the Presbyterian elders quorum were certainly less demanding of time and effort than LDS membership had ever been for me. Our children had regularly attended LDS church services with Janice and now occasionally attended the Presbyterian Church with me. Dan knew a young girl from school who attended my church, and so for some period of time, he preferred to go to church with me rather than his mother. Truth was that neither congregation had at that time flourishing youth programs that could attract and hold the interests of young people.

When the pulpit had been vacant for a while, I was elected in 1992 to serve on the Pulpit Nominating Committee (PNC). There were four of us on this committee of two married ladies and two married men. We began to invite ministers of interest to visit our pulpit for preaching engagements and to look for and interview such people. This process was aided considerably by our attendance at a face-to-face church conference in Dallas, Texas, sponsored by the Synod of the Sun for the express purpose of meeting and interviewing candidates of interest. After reading many dozens of personal information files (PIFs), we had become particularly interested in Larry Tyler. Nevertheless, we interviewed seven prospective ministers, one of them (Larry) twice, before agreeing to further consider three of them. We prayed individually and collectively for God to lead us to the right person. We telephoned numerous credible references, all of whom gave glowing recommendations of Larry and his qualifications for a ministry. We watched and listened to two of Larry's sermons on videotape and heard him in person from the pulpit. We fell in love with his family at first sight. Finally, we concluded with Larry that God had brought us together.

Our dream of filling our pulpit with a pastor who could serve our needs was temporarily shattered when our Presbytery's Committee on Ministry (COM) deemed our candidate to be "unacceptable and not a good match for Thibodaux." At our meeting with them, we asked for further explanation and were told a number of things, including the following: he is a thinking rather than an emotional person; he is inflexible, insensitive, and cannot empathize with people; he is ignorant of church polity; he lacks knowledge of and interest in today's church; he is ignorant of contemporary theology; he is too hard-line on abortion; he just shrugged off some questions from the COM and didn't answer; there are things about him that our Pulpit Nominating Committee (PNC) doesn't know and that we are not privy to.

In spite of the above objections from the COM, the facts about Larry Tyler that we observed and were assured of were: he creates an excitement about God through preaching the word; he has a great knowledge of the Holy Bible and an ability to apply its contents in a meaningful way to life today; he has genuine interest and experience in working with young people; he has a humble, low-key, forgiving, loving, Christlike personality; and his references consistently cited his ability to preach and to work effectively with and relate to all kinds of people on a one-on-one basis.

At the time Larry was being considered for the Thibodaux pulpit, he had three prospective choices of churches in which he might minister. He told one of them "no" while believing that Thibodaux would be the place. Following the sudden last-minute decision of the South Louisiana COM, another COM in Texas quickly ended his third prospect. I came to believe that Larry was treated unfairly by a COM with little if any qualifications to make such decisions. Comparison of our hard-sought conclusions about Larry and those of his numerous references with the above-cited reasons of the COM for disqualifying him led me to inform the COM that: we were entitled to all available information bearing on Larry Tyler's or any other candidate's suitability for our pulpit; Larry should be given another chance before the South Louisiana COM to clear his name and record; and future reservations of the COM regarding pulpit candidates should be discussed with the PNCs involved before final decisions are reached. Receipt of my letter was never acknowledged.

I continued to attend church and to serve on the session, but with less enthusiasm. On May 7, 1995, I addressed a letter to the session and members of the First Presbyterian Church of Thibodaux as follows:

> Dear Brothers and Sisters:
>
> I am writing this letter to ask that you accept with good will my resignation from the session of our church. This has not been a hasty decision and is not related in any way to negative feelings toward members of our congregation or our pastor. I have no such feelings. On the other hand, I have benefited from and am grateful for the hand of fellowship and trust extended to me by this congregation during the last dozen years. I also believe that many good things are happening in the First Presbyterian Church of Thibodaux and that our pastor, Walter Hackney, is an important influence in bringing much of this about.
>
> I and my two brothers were reared in the Presbyterian Church by parents who both were active Presbyterians. My father and his father before him were elders and trustees in the First Presbyterian

Church of Decatur, Alabama. My mother is one of five children reared in a Presbyterian home by a conservative Presbyterian minister and his devout wife. She, all her siblings, and most of her relatives that I have knowledge of have been active, lifelong Presbyterians.

I repeat that my decision has not been a hasty one. I have reached it after careful and prayerful consideration of a combination of facts. I am a pragmatist with little sympathy for tradition. Most Presbyterians feel strong ties with tradition. Since I have no children or immediate family in this church, I am reluctant to push for changes, which would affect whole families of good Christian people who may not agree that such changes are desirable. While I have spent much of my professional life crusading for changes of one kind or another, I do not want to play that role at church under present circumstances.

I cannot, with good conscience, wholeheartedly support our presbytery, synod, or the general assembly of our church. I am strongly opposed to a number of actions taken by one or more of these bodies in recent years. I believe that many of you feel the same way. However, you may have more reasons to stay and "fight" for changes in the bureaucratic church leadership than I. I am the only Presbyterian in my immediate family. My children and grandchildren all are members of other denominations.

While in my thirties and forties, I gave about fifteen years of dedicated service to the Mormon cause before realizing that I could not endorse all of that church's doctrines. My individual needs for worship might be partially satisfied by any one of many different Christian churches. However, my beliefs about a number of things such as church government, organic evolution, and the possible fallibility of scripture may render me undesirable for leadership roles in most of these. My wife is a dedicated Mormon, unwavering in her faith. Her happiness is profoundly and positively influenced when I occasionally attend church with her. Therefore, I have concluded that for the remainder of our natural lives together, I should promote our relationship by attending church with her as often as possible and practicable.

I believe that I shall always be a Christian. However, at least for the present time, I feel that I should have no leadership position in the Presbyterian Church or in any other duly organized church with which I am familiar.

I hope that I may always consider each of you to be my brother or sister of the faith in Jesus Christ and in the family of one Almighty God.

With love and all sincerity,
Henry Long

In spite of this letter, I was attending the Presbyterian Church again regularly in 1997-98, declined an invitation in October 1997 to again serve on the session, and then told Dr. Wes Magee of the nominating committee in August 1998 that I would serve again if called. I also requested that none of my church contributions go to the synod, the presbytery, or the general assembly of the church and that they remain for use by the local congregation. I don't know how God may judge such behavior. I do know that singing with any congregation (Mormon, Protestant, or other) that old familiar hymn, "Onward Christian Soldiers," often brings tears to my eyes and goose bumps to my flesh.

I was interested to learn that a movement to establish churches without creeds, and governed by congregational rule with the Bible as the only official source of guidance, was begun in a significant way by six Presbyterian ministers in 1804 when they wrote and signed the Last Will and Testament of the Springfield Presbytery, which rejected human creeds and stated their will to "sink into union with the Body of Christ at large." This movement led to the present Churches of Christ, Disciples of Christ, and independent Christian churches. In recent years, community churches, with various names and under congregational rule, have become commonplace.

The Mormon (LDS) faith offers much to many believers. Its form of church government is adaptable to endless expansion with seemingly reasonable checks on abuse of authority. Such a worldwide organization or any other, if literally guided by the mind of God, could perform wonders in behalf of mankind. LDS church members recognize that while church leaders may be inspired by God to lead the church, they themselves are entitled to inspiration of the first order in directing their own lives. I have learned by experience to believe that ultimate family welfare should take precedence over church needs when individuals are called to positions of church responsibility. However, active LDS membership, in adequately organized wards of adequate size, offers opportunities for personal development through service to a larger percentage of the membership than may be found anywhere else to my knowledge. One reason for this is that there is no paid ministry, and service of all types is rendered by members without charge. Also, there is an ongoing attempt to insure that as many as possible have church jobs at all times. When the jobs are all filled, the congregation commonly is divided, and the process starts all over again.

A cornerstone of LDS doctrine is that the authority given by Christ to the early apostles was lost through apostasy and reestablished by direct revelation to Joseph Smith in the mid-nineteenth century. While many students of Christian religious history readily admit to a great apostasy throughout the world in the centuries following Christ's life on earth, many also believe that Jesus' promise, "where two or three are gathered together in my name, there am I in the midst of them" (Matthew 18:20), is proof enough that the apostasy was not complete and that there have been at least some small groups of Christians upon the earth continuously since the first century AD. The matter is not subject to unambiguous historical proof one way or the other. Serious Christians generally would love to receive the blessings of new and continuing revelation from God and hope it comes soon. Mormons believe that it has come and is here now. I am presently undecided but have an open mind about it.

A major point of LDS faith that disturbs many other Christians is its emphasis on belief in only one true church with sole authority to act for God, similar in this respect to the Roman Catholic Church. This is not unreasonable to me. For some, such beliefs dwarf the importance of developing a personal relationship with Christ. However, the first line of a favorite LDS hymn reads "I stand all amazed at the love Jesus offers me" and such protestant and LDS favorites as "I Need Thee Every Hour" and "Jesus, Savior, Pilot Me" attest to the importance that Mormons also place upon development of personal relationships with Christ.

Mormon beliefs about God and his church differ significantly in several respects from those held by other Christians. The Mormon practice of proxy baptism for the dead seems strange to many for whom Mormons respond with I Corinthians 15:29: "Otherwise, what do people mean by being baptized on behalf of the dead? If the dead are not raised at all, why are people baptized on their behalf?" What better explanation could there be for this scripture than that baptism for the dead was practiced during New Testament times?

Equally startling to many Christians may be the idea that the likeness of a serpent, as commonly found on artifacts in and around ancient Central American temples, could have any possible relationship to the ancient worship of Christ by people in this area as described in the Book of Mormon. In this regard, some believe it pertinent to recall the biblical scripture "and as Moses lifted up the serpent in the wilderness, so must the Son of man be lifted up, that whosoever believes in him may have eternal life" (John 3:14-15). I would not have made this connection on my own, but it is an interesting thought.

Mormons are best known for their past practice of "polygamy," which was sanctioned and strictly regulated by the LDS Church from 1843 until 1890, when it was officially discontinued. Since 1904 and in order to comply with the laws of our land, church policy has been to excommunicate any member either advocating or practicing plural marriage. Polygamy has been practiced in many cultures at many times and places in which it has served beneficial purposes. Personally, I would prefer no part of it as the responsibilities associated with only one wife are quite sufficient for me. However, the practice was permitted, if not occasionally encouraged, by God among some of his people in Old Testament times and perhaps, for a time, in the nineteenth century as well. Recall Jacob's marriage to four wives, two sisters and their two maids (Genesis 29:16-35, 30:1-26). Also, Moses taught "If brethren dwell together, and one of them die, and have no child . . . her husband's brother shall go in unto her, and take her to him to wife . . ." (Deuteronomy 25:5). It is true that the apostle Paul said "a bishop then must be blameless, the husband of one wife . . ." (1 Timothy 3:2). Perhaps he meant for this to apply also to all members of the church. Of course, Paul also said that women should keep silent in the church (1 Corinthians 14:34-35), which many Christians today do not accept. I believe that young Joseph Smith so antagonized the established Christian community with his claims of direct revelations from God that every "new" idea uttered by him fanned the flames of hatred among many for his new faith. The practice of polygamy as a principle of faith among professing Christian Mormons, was such a shock to the nineteenth century American Christian establishment that it only "added insult to their injury."

The Mormon idea most unpopular and considered blasphemous by many is that: "As man is, God once was; as God is, man may become." I believe that Joseph Smith dealt a significant blow to the growth of Mormonism by pontificating a few weeks before his own assassination, at the funeral service of a man named King Follet, on the subject of the "nature of God." His remarks there, which are now known as "Joseph Smith's King Follet Discourse," apparently gave rise to the above quoted idea, which deals with a subject about which I think that no one knows enough to talk about. A revelation given by Joseph Smith in 1843 states: "The Father has a body of flesh and bones as tangible as man's; the Son also; but the Holy Ghost has not a body of flesh and bones, but is a personage of Spirit. Were it not so, the Holy Ghost could not dwell in us" (Doctrine and Covenants 130:22). In the biblical account of the Creation, we read: "And God said, let us make man in our image, after our likeness . . . So God created man in his own image, in the image of God created he him; male and female created he them" (Genesis 2:26-27). Further desired clarification is not provided by these scriptures. However, I like a statement by Scott Campbell (*Zion's Herald*, November/December 2005, p. 35) who said: "what we don't know about God so far exceeds what we do know that we stand a very good chance of being wrong most of the time."

Recently, I reread a letter, written to me thirty years ago by my friend and stake president, Dr. Gruwell, who presided at my excommunication, in which he briefly reviewed for me, from the Mormon Church's General Handbook # 20, the instructions pertaining to my situation as follows: complete severance from the church; denial of all privileges of the church; not entitled to speak, offer public prayer, partake of the sacrament, or otherwise participate in these meetings; tithing and other contributions are not to be accepted from persons who are excommunicated; individuals are encouraged to repent, attend sacrament, auxiliary meetings, and general conference sessions and to humble themselves through prayer etc. so that they may in due time be reinstated in the church; and all covenants taken in the temple are cancelled.

I have pondered the meaning and significance of doctrine that encourages attendance at sacrament while simultaneously denying partaking of sacrament as punishment for sin. I believe that I was excommunicated primarily for the "sin" of no longer believing that the LDS priesthood has sole authority to act for God. However, Christ said: "This is my body, which is given for you. This do in remembrance of me" (Luke 22:19). And elsewhere we read: "On the first day of the week, when we were gathered together to break bread . . ." (Acts 20:7). In view of these scriptures, it seems that I must choose between following Christ's command to take sacrament "in remembrance of me" or to not take it out of respect for LDS priesthood authority. For me and my present faith, there are many uncertainties. However, I think that the first option is more important. Therefore, I take sacrament these days with my wife.

And now after all that has been said and done, my respect and sympathies for LDS people and their faith are such that I may reunite one day with their church if I do ever find convincing archaeological, historic, or scientific support for the Book of

Mormon story and the claims of Joseph Smith the Prophet. Meanwhile, I continue to pray for enlightenment on the subject.

Faith in the distinguishing beliefs and creeds of many Christian churches goes far beyond the simple belief in an omnipotent God and in the recorded teachings of his son Jesus Christ. I have come to believe that God may not be much concerned about the many variations among church creeds. There is ample ground for varying interpretations of scripture, although the apostle Paul wrote to Timothy to "avoid disputing about words, which does no good" and "have nothing to do with stupid useless controversies; you know that they breed quarrels" (Timothy 2:14, 23-24). Some Christians refrain from arguing about religion while silently feeling superior to or better informed than their misguided or poorly informed neighbors who may adhere to a different variation of the Christian faith. I believe that even such low-profile pride is unjustifiable.

A search of scriptures for the word "pride" gives one a feeling for what God says about that particular vice. The eminent Christian scholar, C. S. Lewis, calls it "the utmost evil" and "the great sin." According to Lewis, "each person's pride is in competition with every one else's . . . Pride gets no pleasure out of having something, only out of having more of it than the next man . . . If everyone else became equally rich or clever or good-looking, there would be nothing to be proud about. It is the comparison that makes you proud: the pleasure of being above the rest . . . the proud man, even when he has got more than he can possibly want, will try to get still more just to assert his power . . . it is Pride which has been the chief cause of misery in every nation and every family since the world began." Even the presumption of pride, where it does not exist, may cause problems and hard feelings.

It seems to me that the matter of major importance is to recognize that "the heavens declare the glory of God; and the firmament showeth his handiwork" (Psalm 19:1). I do not and cannot believe that the universe is the result of random and unpredictable events not planned, not organized, not created by any intelligent being. Whether man was created to worship God or whether he evolved to worship him, the fact is that most men do in some way worship and benefit therefrom. Thousands of years of human existence and modern medical science both testify to these facts.

Many believe "intelligent design" by a creator is a better explanation for the origins of life and of man than is "evolution" and, for this reason, should be given time and space in biological science textbooks. Philosophy and religion deal with beliefs about life and the nature of the universe; and religion is based upon faith, which the Bible says is "the evidence of things not seen." Knowledge in these realms is acquired by reading, studying, and even sometimes by prayer. However, scientific knowledge is acquired not only by study and reason but also by observation to describe or measure. The description of living forms and life processes are the provinces of science. Therefore, "observable phenomena" are essential to science and to distinguish it from other types of knowledge. Inferences made from these studies about a creator are not a part of science since nothing about God can be described or measured objectively. While mathematics might conceivably be mentioned in a history course, little or no time is given it there. Similarly, "intelligent design" might be mentioned in a biological science course, but little if any time can be justified for it there. My belief in God is based upon my own subjective reasoning and experiences.

Principles of evolution and genetics form an important foundation for all life sciences, including agriculture and medicine. Some believe that acceptance of Darwinian evolution would require rejection of their religion. Others believe that the outcome of many random events, on which evolution depends, are at least as predictable for the creator as the probable ratio of fifty-fifty for heads and tails following many fair tosses of a coin. For these good people, reason need not be the enemy of faith.

While a university professor teaching principles of evolution as an essential part of biology, I presented myself as both biologist and Christian, mentioning my church membership, active participation in church work and testifying to my belief in God the creator and in Jesus Christ, my savior. In 1990, a paperback book was written by Dr. Tim M. Berra as "A Basic Guide to the Facts in the Evolution Debate." Unfortunately, the book title is "Evolution and the Myth of Creationism," suggesting a bias that I believe Berra generally avoids in his well-written book. During the last five years of my career in teaching freshman biology, I saw to it that this little book was available in the college bookstore and offered to students the optional chance to raise their final grade by ten points on the basis of a short multiple choice test appended to the final exam and covering Berra's book. Many students took advantage of this opportunity and raised their grades by a letter in doing so. It seemed unfortunate to me that some also cheated themselves of this opportunity by fearing the loss or weakening of their religious faith.

I believe that divinely inspired yet humanly transmitted ideas, including scripture, demand ongoing interpretation. The focus of God's will for mankind, as described in Bible scripture, has always been to love and honor God and to care for those less fortunate than ourselves. The details of how this may best be done necessarily vary in time and place and with the evolution of science and civilization. An example of this is found in the different suggestions from biblical scriptures regarding the roles of women in the church. In some scriptures, Christians are advised that women should keep silent (1Corinthians 14:34-35 and 1Timothy 2:11-12) while in others, women may prophesy and teach (Luke 2:36-38; Acts 2:18, 18:26; Philippians 4:3).

Circumstances in Corinth, where Paul and Timothy both labored for some time, may have differed, in some way pertinent to this matter, from conditions elsewhere. The above scriptures also illustrate the potential error of interpreting selected passages to support an idea without including all scripture possibly relating to the subject in question.

Terrible things often happen to good people and good things to bad people without any apparent reasonable relationship to who is good and who is bad. Saint Augustine wrote that what happens contrary to God's will would not happen if he did not allow it and that he does not allow it unwillingly but willingly in order to bring good out of evil. For those, like me, who believe in an omnipotent being, this thought may sometimes offer comfort. For agnostics and atheists, it may only seem like so much nonsense. However, doubting the existence of a Supreme Being seems like greater nonsense to me.

Today, I am reminded every time I hear, read, or say the Lord's Prayer, "forgive us our sins as we forgive those who sin against us," that God has made it perfectly clear that if we do not forgive, we shall not be forgiven. If I believe these words, I have no alternative but to forgive. I am grateful to C. S. Lewis for his explanation in his book, "Mere Christianity," of what it means to love your neighbor or even your enemy: "we must try to feel about the enemy as we feel about ourselves—to wish that he were not bad, to hope that he may, in this world or another, be cured: in fact, to wish his good. That is what is meant in the Bible by loving him: wishing his good, not feeling fond of him nor saying he is nice when he is not." This advice from God's word is much more than a guide to blessings in the next life. Medical science suggests that it also is a guide to better health here and now.

20

Conclusions

I have written here some conclusions, based on my own experiences, with the hope that they may enlighten or entertain or both, realizing at the same time that "there is nothing new under the sun."

(1) Now as a senior citizen, I see that I have made some progress over the years in personality change but realize that there is yet much room for improvement. Many normal people instinctively know, before reaching adulthood, things that have taken me more than half a century to learn. One may marvel that anyone should require a lifetime to appreciate such basic principles for living.

(2) Writing this book has been a therapeutic experience. The process of recalling past events with associated experiences and fitting these into their proper historical sequence has helped me to see the panorama of my life in a way not before fully appreciated. It has helped me to achieve to some degree that desire expressed long ago by Robert Burns: *"O wad some Pow'r the giftie gie us. To*

see oursels as ithers see us!" In doing so, I am left with the impression that, in many of my life's situations, I have behaved somewhat like a wounded animal, which sometimes fights effectively when feeling threatened, often feels threatened without good reason, and doesn't know how to relax and enjoy when faced with the opportunity.

(3) I have struggled lifelong with problems related to interacting with people, and I believe that the probable shortage of cuddling as a baby, together with a scarcity of early socializing experiences, likely have been root causes of these problems. The earliest influences on living things have the greatest impacts on their lives. Cuddling of baby primates, humans included, promotes self-esteem and has great positive effects on later social behavior. Research has shown that such effects begin even prior to birth since a woman who is stressed while pregnant is more likely to bear a child who will develop an anxiety disorder, such as ADHD or other behavioral problems. Remembering the highly disciplined nature of my dear mother, her desire to be gainfully occupied at every moment of her life, and the difficult times in which she lived her young married life, it is hard to imagine that she ever could have spent much time cuddling her babies or blowing on their belly buttons just to hear them giggle. As the oldest of three brothers, I have no memories of babies being cuddled in our home. My brother's book cited below testifies to the related nature of problems with which he has contended and mostly overcome through much effort and expense.

(4) My particular problems were partially overcome to some degree through years of often prayerful and occupationally required performance in various capacities as a professor, consultant, and church worker. However, in many social situations, I am still uncomfortable and often would rather be reading, exercising at the gym, or working in the yard than sitting at a banquet table, shaking hands at a reception, or mingling at a party. In some professions, such as surgery, I probably would never have been able to keep a steady hand under the scrutiny of witnesses.

(5) Do we ever recover from our childhoods? The right answer may be "sometimes" and "partly," but "never completely." As a senior citizen in my late seventies, I still clearly recall my panic attack at the junior high school oratorical contest. My hands have trembled with emotion many times throughout my life and sometimes to my embarrassment. I have always been subject at times to shortness of breath, sometimes approaching panic in the wee hours of morning when I should be sleeping soundly. Medical exams have shown no physical cause for such symptoms. I have found more relief from physical activity than from medication. Psychiatric treatment, for which I was never willing to consider paying the cost, might have been more effective.

(6) Needham L. Long, MD, in his book, "Panic: An Odyssey—Anger in Disguise" (Dorrance Publishing Co., Pittsburg, PA 2005), describes some causes and effects of severe panic and anxiety attacks and their effective cure by psychiatric treatment. A prominent physician has recommended that this

should be required reading for physicians and medical students. It should be an interesting and informative read for anyone with concerns about anxiety and panic disorders.

(7) From womb to tomb, we crave flattery and success as reflected by the interest of others in what we do. We need to feel good about ourselves for something we have said or done or are able to do. My earliest significant feelings of such success were associated with the impacts of my research on crop production, my activities as a teacher of graduate students, and my experiences as an agricultural consultant. A high school classmate and newspaper reporter once described me by writing "Henry was pleasant, and never one to pick a fight at recess. He just never did much of anything." I was not able to be a serious student until several years later and was never a noted athlete or Casanova. I finally got started giving my best professional efforts to what I thought was most needed and would therefore be most appreciated. With more emotional maturity, I might have realized sooner that other things deserved my best as well.

(8) My evolution from an overly shy and defensive child to a sometimes surprisingly and overly aggressive adult became an apparent enigma. However, it enabled me by determination and hard work to bring about some desirable changes in Louisiana agriculture, although often at the expense of the good will of others. If I had found more confidence earlier in life and developed a friendlier personality to go with other abilities, I might have accomplished more. It has been jokingly said by

some that "to be a productive research scientist, one must have a defective personality."

(9) Many positive benefits of hard work are lost when one's work becomes a place to hide or an excuse to avoid participating in social activities. From 1949, when I began to take college studies seriously, until 1999, when I retired from breadwinning activities, I was the ultimate workaholic. Work was not only preparation for survival and a means of survival, but it also became a place to hide and an excuse for avoiding many public and social engagements. In this process, I learned to find joy in the planning and accomplishment of a job well done but often missed out on other normally associated benefits.

(10) Society generally rewards one for contributions in proportion to how well one explains their significance to maximum numbers of the right people. To do this, one must write and speak well enough and must be able to effectively transmit his or her message under appropriate circumstances, including official or professional as well as social situations. If this is left to others, then others will reap the benefits from your efforts and expertise that could and should have belonged to you.

(11) Religious character requires faith by the exercise of which great and unique benefits are derived. Although this demands the placing of trust in beliefs not proven or provable by science, the benefits are worth the efforts and include: (1) psychological comfort in times of need, (2) the sense of belonging to a recognized culture group with attendant social benefits, (3) enhanced opportunities to develop social skills, and (4) opportunities to learn

and practice principles for living as taught by Jesus Christ. The central admonition of Christ to "love God and thy neighbor as thyself" addresses some of the greatest human needs and may be the only plan in sight by which mankind may ultimately avoid self-destruction and reasonably hope for universal well-being. Loving both God and neighbor should improve with practice. Major purposes of church attendance should be to help insure that these goals are met.

(12) When offences occur, happiness cannot be restored by repentance to God or church without sincere expressions of sorrow to the offended for suffering inflicted. In his Sermon on the Mount, Christ said: "If you are offering your gift at the altar and there remember that your brother has something against you, leave your gift there before the altar and go; first be reconciled to your brother and then come and offer your gift" (Matthew 5:23-24). As it is with one's brother or sister, mother or father, daughter or son, so it is important with one's spouse.

(13) A lack of agreement on how money will be used may become a major problem in marriage. It is difficult to imagine the many possible ramifications of these disagreements until actually faced with them. The basic problem hinges upon the couple's beliefs and feelings relative to the importance of "immediate gratification" versus "future security." Young couples would do well to discuss this frequently and thoroughly when marriage is their goal, even at the expense of time for more enjoyable activities.

(14) Serious offenses may be forgiven, but may never be forgotten, and the memories always hurt.

However, such hurts may become useful reminders. The midlife crisis in my marriage would never have occurred if I had been a more attentive and less neglectful husband, more loving and less emotionally abusive, more caring and less like a worker ant.

(15) In romance and in marriage, as in other life experiences, first impressions are of great importance. "When a man is newly married, he shall not go out with the army or be charged with any business; he shall be free at home one year, to be happy with his wife whom he has taken" (Deuteronomy 24:5). Many broken and damaged marriages begin with too much work and worry and too little happiness. Men should always give serious thought to how they may best project their feelings of love, joy, and respect for their sweethearts, particularly in their first moments of greatest intimacy.

(16) The continuous setting of goals, pursued relentlessly, achieves much in spite of obstacles and shortcomings. I made significant contributions through research, teaching, agricultural consulting, foreign assignments, and, in some small ways, through church work. In the course of these activities, I learned to interact with others adequately perhaps, and sometimes admirably, while acting in official and professional roles. I even learned a little, but not enough, about relaxing and enjoying social interactions. Perhaps if I should live another hundred years, I might achieve a normal and healthy personality.

(17) Fame is short-lived and fortune commonly depends on profiting from the efforts of others. Profiting from

the work of others requires effective delegation of responsibility. I was always uneasy delegating anything to more than a very few. I liked to think that scouting a farmer's fields, upon which his livelihood depended, was responsibility akin to prescribing treatment for life-threatening disease. If everyone was like me in this respect, the largest corporations in the world would have no more employees than the fingers on one hand. Considering my lack of business and people skills, I was fortunate to have kept a profitable service business going for thirty-five years.

(18) People may forget what you said and what you did but will never forget how you made them feel. I cannot claim credit for discovering such wisdom. This was paraphrased from an e-mail letter sent by a friend. However, its truth is recognized immediately by most who have lived long enough to do so. I regret that my experience and training did not focus more on making others feel good. I would have benefited much from a larger serving of these common skills of politicians and salesmen.

(19) Outstanding among my regrets are missed opportunities to create more happy memories and to spend more personal and individual time with family. My dear wife tried hard to make up for these deficiencies and succeeded to a great extent. But she was handicapped by the frequent absence of one who should have been a more valuable member of the home team. When my absence was not actually physical, it was often just as real due to my focused attention on matters of responsibility unrelated to family.

(20) My wife long suffered in a variety of ways from the misfortune of marrying a man so emotionally crippled, so insecure, and so preoccupied with trying to heal himself and with climbing mountains to prove himself that he was unable to give her the love and attention she deserved during the first twenty years of their marriage.

(21) Although separated geographically and ideologically from some of them, I love all of my children and grandchildren and pray regularly for their happiness and welfare.

(22) A wise person has said, "the great essentials of happiness are something to do, someone or something to love, and something to hope for." This may be the outline for a good plan, but filling in the details and implementing the program will require much more than good intentions!

Summary

This true story begins with the early experiences of a young boy during the Great Depression whose parents with their three sons moved from town life to primitive conditions on a small farm where they could produce much of their own food, teach their children to work hard, and avoid some of the corrupting influences of society. The boys all learned to work hard; but due to an unusual combination of hard times, isolation from peers, and a strictly disciplined puritanical upbringing, they missed out a great deal on learning early the skills of social interaction. These circumstances led to low self-esteem, frequent anxiety, and sometimes panic, which the author can remember from about age six or seven and which mostly intensified thereafter through his middle teens.

Following high school graduation, a wasted year at college, and a three-year enlistment in the air force, he began to grapple with life in the real world—quite a learning experience. In spite of having always been a shy and mediocre student in the grade and high schools, he finally learned to study and became an honor student, eventually earning a PhD degree in entomology. By this time, he is also married, and the first of four children has arrived on the scene. In spite of these experiences, Dr. Long remained, in many ways, a social cripple with abnormal fears of public speaking and even casual social contact.

In his first full-time employment as an entomologist at LSU, he became the ultimate workaholic using his work as a place to hide from social activity until partially forced from his hiding place by conversion to his wife's faith and activity in her church (LDS or Mormon). Both church activity and employment, combined with his wife's civilizing influence, forced him to conquer some of his hang-ups about public speaking and social interaction. At LSU, his involvement in research and the dissemination of results from his and others' work embroiled him in a controversy lasting several years, but one which drastically changed sugar cane insect pest management practices for the better in spite of strong and bitter opposition. The positive recognition from these activities gave him a much-needed boost of self-confidence, a previously rare "commodity" in his life. He thirsted for more of the same.

After eight years and advancing through the faculty ranks at LSU, he obtained a faculty appointment at a small regional university enabling him to start a consulting service on the side, thereby significantly increasing his family's income. The painful transition from teaching graduate students to that of college freshmen was gradually accomplished. He also encountered frustrations of faculty in this small university with questionable leadership. Some unfortunate business decisions and acceptance of a short-term United Nations assignment to Egypt separated the family and further aggravated an already-troubled marriage to the point of planned permanent separation. This midlife crisis is resolved over a period of months by faith, love, children, and the realities of twenty years already invested in their family, which reunited and moved to Brazil for a year where Dr. Long completed another UN assignment.

Upon returning to Louisiana, he is faced by business and professional challenges that he deals with in a manner often involving personal confrontations. Before retiring, he was officially honored for pioneering, developing, and practicing the environmentally sound practice of "integrated pest management" of sugar cane insects. Following retirement, he was elected to the Louisiana Agriculture Hall of Fame. Because of confrontations in his religious environment, he evolved from Presbyterian to Mormon to Presbyterian and eventually nondenominational Christian. Now retired and living in Tennessee, he discusses many conclusions he has drawn from a long and confrontational life.

Author Biography

William Henry Long III was born September 20, 1928, in Decatur, Alabama. He attended public schools, graduating from Decatur High School in 1945. After a year of college and a three-year enlistment in the air force, he earned the BA degree in zoology and entomology from the University of Tennessee at Knoxville in 1952, followed by MS and PhD degrees in entomology, respectively, from North Carolina State University at Raleigh in 1954 and Iowa State University at Ames in 1957. While pursuing graduate studies at North Carolina State, he met and married Janice Helon Rogers with whom he has four children, eleven grandchildren, and two great-grandchildren at last count.

He was employed in the Louisiana State University and Agricultural Experiment Station at Baton Rouge as assistant entomologist (1957-59), associate professor (1959-64), and professor (1964-65) before accepting appointment as professor of biological sciences (1965-85) and later distinguished service professor (1985-94) at Nicholls State University in Thibodaux, Louisiana, until his retirement from university service in 1994. Upon leaving LSU in 1965, he also started an agricultural consulting service, later known as Long Pest Management Inc. in which he worked, continuing after back surgery in 1996 until health problems and other factors forced his retirement in 1999.

During several short intervals, he took leaves from Nicholls to serve United Nations appointments with: (1) the Food & Agriculture Organization (FAO) as entomological consultant to the Ministry of Agriculture, Cairo, Egypt, October 1973 until January 1974; (2) the International Atomic Energy Agency (IAEA) as an expert in entomology to the Center for Nuclear Energy in Agriculture (CENA), University of Sao Paulo, Piracicaba, Sao Paulo, Brazil, January to April 1975 and October 1975 to November 1976.

His responsibilities at LSU were primarily in research on the biology and control of sugar cane insects. However, he also served as chairman of ten graduate committees for MS and PhD candidates in entomology during eight years there and taught a graduate course in the fundamentals of applied entomology during the last three of those years. At Nicholls State, he taught a general course in entomology each fall semester and animal biology for freshmen students every semester for many years. His contributions through research, publication, teaching, and consulting have been recognized by the appearance of his biographies in American Men and Women of Science (editions 10-17) in Outstanding Educators of America (1971) and by his election to the Louisiana Agriculture Hall of Fame (2000). Also, in Miami, Florida, at the 1993 Inter-American Sugar Cane Seminar on "Sugar Cane and Our Environment," he was honored and presented a plaque of recognition "in appreciation of his outstanding work in pioneering, developing, and practicing the environmentally sound practice of Integrated Pest Management."

He was a member of the American Society of Sugar Cane Technologists, the American Sugar Cane League, the Entomological Society of America, the Louisiana Entomological Society (president, 1970), the Louisiana Ag Consultants' Association (president, 1977), and the National Alliance of Independent Crop Consultants (NAICC). He held three board certifications by the Entomological Society of America, was a certified independent crop consultant and researcher by the NAICC, and an agricultural consultant certified in several fields by the Louisiana Department of Agriculture.

He now lives with his wife in retirement in Tennessee, near one daughter, a son-in-law, and three grandchildren.